Architecting Spacecraft with SysML

A Model-based Systems Engineering Approach

SANFORD FRIEDENTHAL

CHRISTOPHER OSTER

Copyright © 2017 Friedenthal & Oster

All rights reserved.

ISBN: 1544288069
ISBN-13: 978-1544288062

CONTENTS

Preface		i
Acknowledgements		ii
Chapter 1	Introduction	1
Chapter 2	SysML Overview	5
Chapter 3	MBSE Method Overview	25
Chapter 4	Plan the Modeling Effort	29
Chapter 5	Setup the Model	33
Chapter 6	Analyze Mission & Stakeholder Needs	39
Chapter 7	Specify Spacecraft Requirements	57
Chapter 8	Synthesize Alternative Architectures	69
Chapter 9	Perform Design Analysis	107
Chapter 10	Manage Requirements Traceability	115
Chapter 11	Verify System	119
Chapter 12	Summary	125
Appendix A	References	127
Appendix B	About the Authors	129

PREFACE

This book provides an example of how a Model-based Systems Engineering (MBSE) approach using an MBSE method to perform the systems engineering activities and the Systems Modeling Language (SysML®) to model the system can be applied to the architecture design of a spacecraft. This approach focuses on developing a system model to support the flow down of mission requirements to the spacecraft and its components. The model is used to identity and evaluate alternative spacecraft architectures, and select a preferred system design to satisfy the mission and system requirements.

Although this is a spacecraft example, the MBSE approach and the system model can be used as an example for application to other domains as well. The system model of the spacecraft used in this book is available at http://www.sysml-models.com/spacecraft, and can be accessed in HTML, or as the native mdzip file that was created using the No Magic Cameo Systems Modeler® (CSM) tool. The approach can be similarly applied using other system modeling tools.

This book can be used as an introduction to MBSE, and provides an example of how to model a system using SysML. It also can be used as a template for documenting an organizations MBSE approach. This book is intended to complement other references that describe the SysML language in more detail.

ACKNOWLEDGMENTS

The authors wish to thank the editors, John Hsu and Richard Curran, and the publisher (AIAA) for publishing the original content of this book as Chapter 4 of 'Advances in Systems Engineering' [1]. The authors would also like to recognize the authors of the New SMAD [2] for the domain specific knowledge and systematic treatise on spacecraft mission analysis and design that was used to inspire the example in this book. In addition, the authors would like to thank Todd Bayer from NASA Jet Propulsion Lab and Jamie Kanyok from Lockheed Martin Corporation, for their in-depth review and excellent insights. The authors also extend a special thanks to Kat Hanna for designing the cover of this book.

Finally, the authors would like to acknowledge the following tool vendors for the use of their software, and the Object Management Group (OMG) as the organization that provides the SysML® specification [6].

No Magic Cameo Systems Modeler®, AGI Systems Tool Kit®, Microsoft Excel®

INTRODUCTION 1

This book demonstrates how a model-based systems engineering (MBSE) approach is applied to develop a preferred system architecture of a small spacecraft. The example used to demonstrate the approach is inspired by the spacecraft design from the FireSat II example in Space Mission Engineering: The New SMAD [2]. The MBSE approach uses an MBSE method to perform the systems engineering activities and the Object Management Group (OMG) Systems Modeling Language™ (SysML®) [6, 7] supported by a modeling tool to model the system.

The book begins with an overview of what MBSE is and why we do it. It explains how an MBSE approach contrasts with a more traditional document-based systems engineering approach to highlight the potential value that MBSE contributes to the design of a system. Chapter 2, then introduces SysML as a graphical modeling language for representing systems. Chapter 3 introduces an MBSE method that describes how to perform systems engineering activities to produce an integrated model of the system. Chapters 4-11 describes the application of this MBSE method using SysML to the development of a preferred system architecture of a small spacecraft, including the flow down of mission requirements to the spacecraft and its components. Chapter 12 provides a summary and parting thoughts.

The emphasis for this book is on the spacecraft architecture and the identification and selection of the key components of the spacecraft, and not on the detailed design of the components. The term reference architecture is used to refer to a more general architecture that can be realized by multiple design configurations. In this book, the reference architecture is used as a basis for performing trade studies among alternative spacecraft architectures to satisfy the mission and stakeholder needs. A similar approach can be used to design a family of spacecraft that span multiple missions such as the CubeSat Reference Model [3].

The term spacecraft is used throughout this book, which refers to any engineered vehicle that is intended to travel in space beyond the Earth's atmosphere. A satellite is a particular kind of spacecraft or natural object that orbits the Earth or any other celestial body.

1.1 MBSE Overview

Model-based systems engineering (MBSE) is an approach to systems engineering where the model of the system is a primary artifact of the systems engineering process. The system model describes the system in terms of its specification, design, analysis, and verification information. More detailed design information is captured in many other engineering models. This system model is managed and controlled throughout system lifecycle to provide a more consistent and precise definition of the system that is traceable to the mission requirements.

This MBSE approach is often contrasted with a more traditional document-based approach where information about the system is captured in document-based artifacts, such as text documents, spreadsheets, and drawings. These documents are managed and controlled throughout the system lifecycle. However, it is difficult to maintain consistency, precision and traceability when the information is distributed across these different artifacts.

In a MBSE approach, the use of a system model can provide significant benefits over the document-based approach that include improvements in the system specification and design quality, improvements in productivity, and enabling a shared understanding of the system. In particular, an MBSE approach can:

1. Increase precision of the system specification and design resulting in reduced downstream errors
2. Improve traceability between system requirements, design, analysis, and verification information to enhance system design integrity
3. Improve the ability to maintain and evolve the system specification and design baseline throughout the system lifecycle
4. Support reuse across projects
5. Provide a shared understanding of the system to reduce miscommunication among the development team and other stakeholders

MBSE is not new, but its emphasis has changed over the years. Systems engineering has traditionally included the use of many kinds of models. However, the current emphasis for MBSE is on building an integrated system model that provides multiple views of the system. The evolving modeling standards, such as SysML, are enabling MBSE to mature and achieve more widespread adoption within the Aerospace and Defense industry and other industries.

Model-based approaches are pervasive in many other disciplines such as mechanical design, electrical design, control systems design, and software design. MBSE is often placed in the broader context of model-based engineering (MBE) [4]. In this context, the system model is intended to integrate with other models used by systems engineers and other disciplines. The system model serves as an integrating framework for other models that span different disciplines. In this way, MBSE enables MBE.

1.2 Kinds of Models

A **model** is a physical, mathematical, or otherwise logical representation of a system, entity, phenomenon, or process [5]. A model describes various aspects of a system. For example, a three dimensional **geometric model** is created from a computer-aided design (CAD) tool to represent the geometrical layout of a system. An **analytical model** of a system may describe quality characteristics of a system, such as its reliability using closed form equations, or the dynamic behavior of a system using **simulations** that provide discrete-time numerical solutions to differential equations.

A model should be developed to enable understanding of some aspect of the system, entity, phenomenon, or process being modeled. The breadth, depth, and fidelity of this model may vary depending on its purpose. For example, a low fidelity geometric model may be sufficient to support trade study analysis during a conceptual design phase. Another simple example is a spring-mass model where the spring constant and first order damping coefficient may be a sufficient representation to answer certain questions, whereas second order affects such as velocity coefficients may be necessary to understand the more detailed dynamics.

A **system model** as used in this book can be referred to as a model that describe logical relationships between different aspects of the system and its environment. A simple example is a block diagram that describes the interconnection between the system components. This is neither a geometric model that describes the system geometry, nor is it an analytical model that produces numerical results. However, it describes important aspects of the system. A system model can be used to assess various qualities of the system, such as its interface compatibility with other systems or between its components, and traceability between the system design and its requirements.

To fully benefit from a shared system model among the development team and other stakeholders, this model must integrate with many other kinds of models and data sources. In the example above, the mechanical components represented in a system model can be related to the same components represented in a CAD model. Similarly, key parameters in the system model, such as selected performance parameters, can be related to the corresponding parameters in other analytical models. Typically, one of the models is designated as the authoritative source for a selected data set at a given point during development, such that when the data in the source model changes, any model with data that depends on the source model may need to change as well. The source model for that data set may change as the design matures. The correspondence between shared information in different models must be maintained along with the identification of the source model for each data set. Model integration is the focus of many industry standards and related activities.

CHAPTER 2

SYSML OVERVIEW

The **OMG Systems Modeling Language™** (SysML®) [6] is a general purpose graphical modeling language for representing systems. It is intended to enable an MBSE approach to support the specification, design, analysis, and verification of systems that may include hardware, software, data, personnel, procedures, and facilities. **SysML** is an extension of the Unified Modeling Language (**UML**) that was adopted by the Object Management Group (OMG) and formally released as SysML v1.0 in 2007. The current version of SysML, as of the time of this writing is SysML v1.5, but the language continues to evolve through the OMG technology adoption process. SysML is also an international standard [7].

This chapter provides a brief introduction to SysML concepts. There are several other books [8, 9] that describe the SysML concepts in detail,. .

SysML can be used to capture multiple aspects of the system including its requirements, structure, behavior, and parametric relationships. These are often called the 4 pillars of SysML. Specifically, SysML can be used to describe the following:

1. the system breakdown as a hierarchy of subsystems and components
2. the interconnection between systems, subsystems, and components
3. the behavior of the system and its components in terms of the actions they perform, and their inputs, outputs, and control flows
4. the behavior of the system in terms of a sequence of message exchanges between its parts
5. the behavior of the system and its components in terms of their states and transitions
6. the properties of the system and its components, and the parametric relationships between them
7. The text-based requirements which specify the mission, system, and components, and their traceability relationships to other requirements, design, analysis, and verification

The following section summarizes some of the key SysML concepts to help understand the spacecraft model discussed in the remaining chapters. The SysML modeling terms are bolded when they are first introduced, and the terms that refer to model elements shown in the diagrams are in italics.

2.1 SysML Diagrams

The nine kinds of **SysML diagrams** are shown in Figure 2.1. These diagrams are used to present the different aspects of a system as described above. The diagrams include the *Requirement Diagram*, 4 kinds of *Behavior Diagrams*, 2 kinds of *Structure Diagrams*, the *Parametric Diagram*, and the *Package Diagram*.

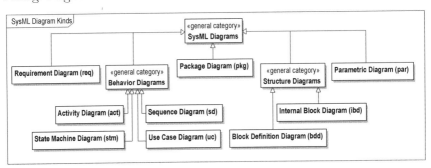

FIGURE 2.1
The nine kinds of SysML diagrams.

Each SysML diagram has a **diagram frame**, a **diagram header** and a content area as shown in Figure 2.2. The frame is the enclosure for the diagram content. The diagram header is shown at the top left of the diagram and describes information about the diagram. The header includes the following four fields although the display of each field is not required:

1. Field 1: short descriptor identifying the kind of diagram, such as bdd for block definition diagram.
2. Fields 2 and 3: kind of model element and the name of model element that the frame represents.
3. Field 4: name of the diagram to indicate the purpose of the diagram.

FIGURE 2.2
Each SysML diagram has a diagram frame with a diagram header that can display fields which include the kind of diagram and the diagram name.

2.2 Modeling Structure

The two diagrams used to model structure in SysML include the block definition diagram and the internal block diagram. The block definition diagram is typically used to define the structural elements called blocks and relate one block to another using various relationships, such as whole-part. The internal block diagram defines how the parts are interconnected.

Block definition diagram. SysML includes the concept of a **block** as a general purpose construct to represent a system, external system, or component. The component may be hardware, software, data, facility, or a person. More generally, the block can represent any logical or physical entity, including things that flow, such as water or information.

A block contains **features** that can include representation of its properties, functions, interfaces, and states. A simple example of a block is shown in the **block definition diagram** (bdd) in Figure 2.3. In the figure, the *Spacecraft* is a block that has a *mass* of 150 kilograms and performs a function called *collect observation data*. It also has a port called *lv electrical i/f* that provides the electrical interface to the Launch Vehicle. The *Spacecraft* can be specified to have many other properties, functions, and interfaces, as well as other kinds of features. Each kind of feature can be depicted in a unique block compartment.

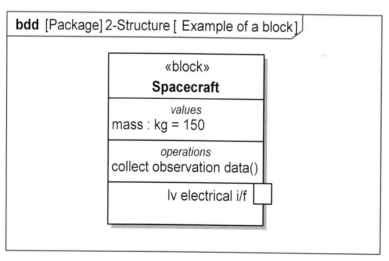

FIGURE 2.3

An example of a block with some of its features.

The diagram header in this figure designates the kind of diagram as a bdd, which is short for block definition diagram. The name of the diagram is called *Example of a block*. The 2nd and 3rd fields in the diagram header indicate that the diagram frame represents the *Package* called *2-Structure*. Packages are explained in Section 2.6. Some of the fields in the header are not displayed in many of the remaining diagrams to simplify the diagrams.

Relationships between blocks. The block can be related to other blocks using whole-part, reference, generalization/specialization, and connection relationships.

A **whole-part relationship**, also known as a **composite relationship**, defines a hierarchy of blocks. A partial *Spacecraft* decomposition into subsystems is shown in Figure 2.4. The *Spacecraft* in this diagram is the same *Spacecraft* that was shown in the previous diagram, but its features are not shown. The black diamond designates the *Spacecraft* as the whole end of the whole-part relationships, and the arrow designates the subsystems at the part ends of the whole-part relationships. Each subsystem is designated with the key word «*subsystem*».

The Guidance, Navigation, & Control (*GN&C*) *Subsystem* is further decomposed into a *Reaction Wheel* and *GN&C SW*. The key words designate these elements as hardware and software components.

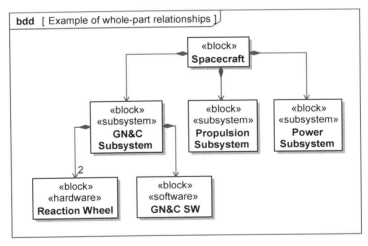

FIGURE 2.4
Partial decomposition of Spacecraft into its subsystems.

Multiplicity. On the block diagram in Figure 2.4, the *GN&C Subsystem* is composed of 2 *Reaction Wheels*, as indicated by the number 2 next to the part end of the whole-part relationship. This is called the **multiplicity**. More generally, the multiplicity has a lower bound and upper bound, such as 2..4, which refers to a lower bound of 2 and upper bound of 4. An unlimited upper bound is referred to with an asterisk (*). A lower bound of 0 means the component part is optionally included as a part of the whole. If no multiplicity is shown on the part end, the default multiplicity is one.

Referencing a block. It is sometimes useful to aggregate a set of components into a logical whole even though the component may already be part of another whole-part relationship. For example, a software component may be logically aggregated into a *Software Subsystem* even though it is already part of another subsystem. As shown in Figure 2.5, the Software Subsystem logically aggregates the *GN&C SW* and the *Power Mgmt SW* using a **reference association** designated by a white diamond on the whole end. This is contrasted with the black diamond indicating these same software components are part of the *GN&C Subsystem* and *Power Subsystem*. The black diamond indicates the primary decomposition of the system. This enables the same components to be aggregated in different ways for different purposes.

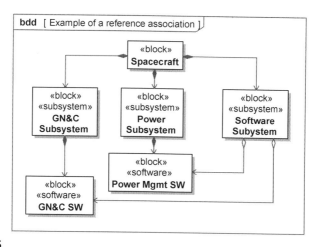

FIGURE 2.5

The Software Subsystem logically aggregates the software components using the reference association.

Generalization/specialization relationship. Another SysML relationship between blocks is the **specialization relationship**. In particular, a more general block contains features, such as properties, functions, and interfaces that a more specialized block can **inherit**. The more specialized block can then add its unique features. This avoids having to define the common features for each specialization, and thereby facilitates reuse.

A simple example is shown in Figure 2.6 showing the *Propulsion Subsystem* having two specializations to represent a *Mono-Propellant Propulsion* and *Bi-Propellant Propulsion Subsystem* variants. Both specializations inherit the *mass* and *fuelMass* properties as designated by the carrot symbol (^). In addition, the *Bi-Propellant Propulsion Subsystem* adds a unique property to specify *oxidizerMass*.

The specialized subsystems can also **redefine** features of the more general subsystem. In this example, the more general *Propulsion* has 1 to 2 tanks, but the *Mono-Propellant Propulsion Subsystem* has 1 tank and the *Bi-Propellant Propulsion Subsystem* has 2 tanks, as shown in their parts compartment.

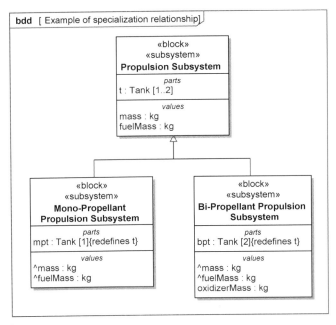

FIGURE 2.6

The Mono-Propellant Propulsion and Bi-Propellant Propulsion Subsystems inherit common features from the more general Propulsion Subsystem, and also redefine the number of tanks.

Internal block diagram. The system, subsystems, and components can be interconnected on an **internal block diagram** (ibd) as shown in the simplified example in Figure 2.7. This shows the connection and the flow of *Electrical Power* from the *Launch Vehicle* to the *Spacecraft*. This connection is further delegated from the port on the *Spacecraft* to the port on the *Power Subsystem*, which distributes the power to other subsystems such as the *GN&C Subsystem*.

ARCHITECTING SPACECRAFT WITH SYSML

The general name for an interconnected element is a **part**. A **port** is an interaction point on a block or part that enables its interface to be specified. The lines between the parts are called connectors. A **connector** can connect parts directly, or to ports on the parts.

The frame of the diagram represents the higher level block that composes the *Spacecraft* and *Launch Vehicle*. In this case, the higher level block is called the *Mission Context* block as shown in the diagram header. When a part is shown as a dashed rectangle instead of a solid rectangle, then this part is referenced by the higher level block.

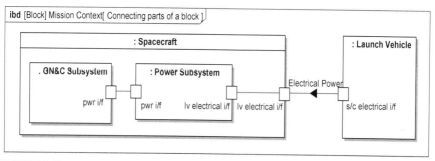

FIGURE 2.7

The internal block diagram shows the connection and flow of Electrical Power between the Launch Vehicle, Spacecraft and some of its subsystems.

Definition versus usage. SysML facilitates model reuse by distinguishing the definition of an entity, and the usage of an entity in a particular context. In Figure 2.8, the *Avionics Subsystem* is composed of a primary and backup computer. Both computers have the same definition but are identified as different usages in the context of the *Avionics Subsystem*. A usage in SysML is a part, and a definition is a block. In this example, *primary* and *backup* are parts (i.e., usages), and the *On-board Computer* is the block (i.e., definition). Both the primary and backup parts are defined (i.e., typed) by the same block.

The two parts are shown in the internal block diagram in Figure 2.9. The colon notation is used to designate the two parts and their type as *primary:On-board* Computer and *backup:On-board* Computer.

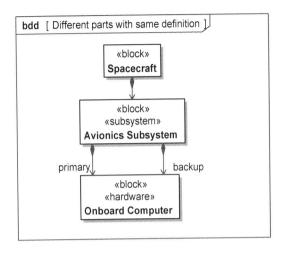

FIGURE 2.8

The parts (primary and backup) represent different roles with the same definition (On-Board Computer).

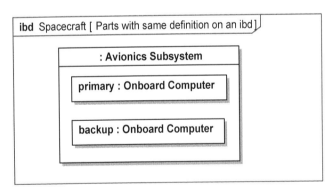

FIGURE 2.9

The parts are shown on an internal block diagram using the colon notation as *partName:blockName*.

2.3 Modeling Behavior

There are 3 different kinds of behavior representations in SysML including activity diagrams, sequence diagrams, and state machine diagrams. Use case diagrams are often used to support behavior modeling as well. Each of these diagrams are summarized below.

Activity diagram. Activity diagrams are used to model input, output, and control flow. An **activity** transforms a set of inputs to outputs through a controlled sequence of **actions**. Figure 2.10 shows a portion of an **activity diagram** to *Provide Electrical Power*. The enclosing frame represents the activity which contains actions. Each action is performed by a component of the *Power Subsystem* that is designated by a **swim lane** (aka **activity partition**). Inputs and outputs to and from the activity are designated as rectangles on the frame, and the inputs and outputs to and from each action are shown as small rectangles on the action. The output of one action connects to the input of another action through an **object flow**.

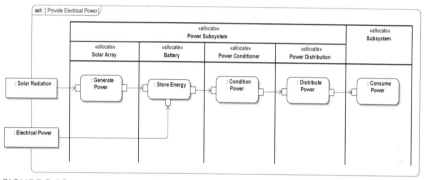

FIGURE 2.10

An activity diagram that shows the inputs and the actions performed by the components of the Power Subsystem to Provide Electrical Power.

The Perform Mission activity shown in Figure 2.11 represents a typical mission scenario. The **control flows** are indicated by dashed lines with no rectangles on either end.

The activity execution semantics are based on the flow of **tokens**. An action cannot execute until a token is available on all of its inputs. This enables the modeler to precisely specify a system behavior in terms of when an action executes, and what inputs and output tokens are consumed and produced.

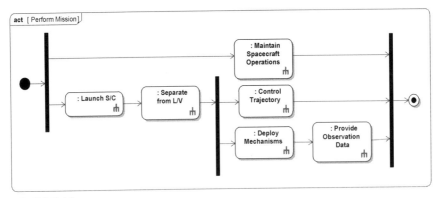

FIGURE 2.11

Activity diagram showing the actions and control flow to Perform Mission.

There are several kinds of **control nodes** to control the flow tokens through an activity. These include the **initial node** and the **activity final node** that designate the start and termination of the activity as shown in Figure 2.11. When a token is available on the initial node, the activity begins execution, and when a token reaches the activity final, the activity terminates its execution.

There are other control nodes such as a **decision node** represented by a diamond that designate which outgoing flow executes based on evaluation of a **guard condition**, and a **join node** that cannot execute until a token has arrived on each incoming flow (refer to join node following :Provide Observation Data in Figure 2.11). The tokens can be control tokens or object tokens that represent things that flow.

There are many specialized kinds of actions. A signal can be sent using a **send signal action**, and a signal be received using an **accept event action**. The sending and receiving of the signal correspond to events that can control the flow of execution. A **call behavior action** (refer to actions in Figure 2.11) can also call another activity which is further elaborated by other actions. This can be considered a form of activity decomposition that can be presented as a hierarchy of activities similar to a traditional functional decomposition. Activity diagrams include many other features to precisely specify a behavior, such as the ability to interrupt selected actions based on the arrival of a signal, and **streaming** and time-continuous inputs and outputs.

Sequence diagram. A **sequence diagram** is used to describe behavior as a sequence of **messages** between parts. It can also represent a timeline of events, such as the simplified mission timeline shown in Figure 2.12. The parts are represented as **lifelines** at the top of the figure, and the messages are sent from one part to another as indicated by the lines with arrowheads.

Time advances down the vertical axis. In this simplified timeline, *Launch Ops Facility* sends a *Power On* signal to the *Spacecraft*, and then sends a *Launch* signal to the *Launch Vehicle*. The time between the *Spacecraft Separation Complete* event and the *Solar Array Deployed* event is constrained to occur between *tmin* and *tmax*.

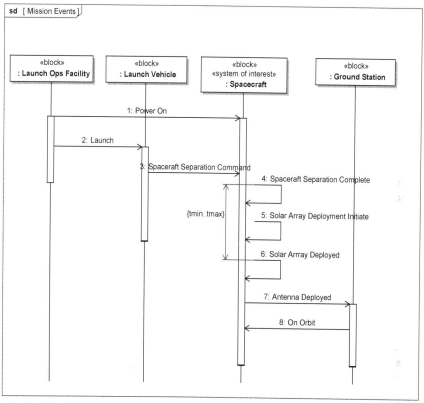

FIGURE 2.12

A sequence diagram used to represent a mission timeline.

A message can represent a **signal** that is sent from one part to another, or a request from a sending part to a receiving part to perform some behavior, similar to subroutine call in software. Sequence diagrams are often used to represent the interaction between software components, but can also be used to represent system level behavior that emphasize message exchange and event sequences. Activities are often used to model continuous behavior, and behaviors involving more complex flows of data and control.

State machine diagram. A **state machine** diagram describes the discrete **states** of a block, and the **transitions** from one state to another. A state represents a condition of a block, such as the off state or on state of a system or component. The transition between the off and on states may be **triggered** by the receipt of a signal, such as when a user turns the power on or off. A simple example of a state machine diagram for the *Payload Sensor* is shown in Figure 2.13.

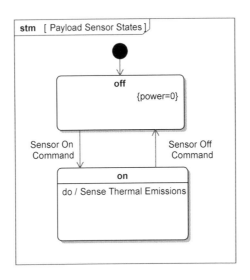

FIGURE 2.13

State machine diagram showing the on and off states of the Payload Sensor, and the behavior called Sense Thermal Emissions that is performed while in the on state.

In addition to defining the states, one can also define **entry**, **exit**, and **do behaviors** that occur when the system or component is within a particular state. These represent the behavior that occurs upon entry to the state, upon exit from the state, or while in the state, respectively. The *Payload Sensor* has a do behavior in its on state called *Sense Thermal Emissions*.

Behaviors can also be defined on transitions between states. For example, a behavior on a transition to send a signal can trigger a state transition in another block. The entry, exit, do, and transition behaviors can be elaborated by activity diagrams or sequence diagrams, or with behaviors that are specified in code called **opaque behaviors**.

With an appropriate MBSE method, SysML can be used to integrate activities, state machines, and sequence diagrams to emphasize different aspects of the system behavior.

Use case diagram. A **use case diagram** is generally used to describe the goals of a system, such as those associated with the mission objectives in Figure 2.14. They consist of the **use case** (e.g. the goal), the system called the **subject**, and the external systems called **actors**. The subject interacts with the actors to achieve the goal. The actors in this figure include the *Forest Service*, the *Operator*, and the *Fire Department*. A goal of the *Operator* is to *Provide Forest Fire Data in Near Real Time* to the *Fire Department*. This supports the broader goal of the *Forest Service* to *Detect and Monitor Forest Fires in US and Canada* as indicated by the include relationship. The actors can be depicted as either stick figures or rectangles with the key word «actor».

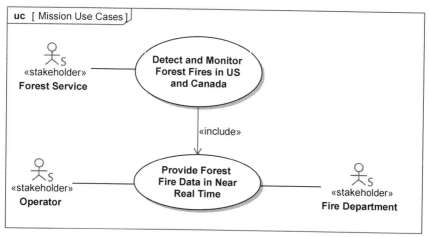

FIGURE 2.14

A use case diagram depicts the goals of the external users (i.e., actors) that the system is intended to support.

Once a use case is defined, activity diagrams, sequence diagrams, and/or state machine diagrams are used to elaborate the behavior needed to realize the goals.

2.4 Modeling Constraints

Constraint block. As noted in the structural modeling in Section 2.2, a block can contain **value properties** that represent important quantifiable characteristics about the system or components, such as its mass, power consumption, size, reliability, or cost. In addition to capturing the key properties of a system, SysML enables the capture of **constraints**, such as Newton's Law. Constraints are captured as **constraint blocks** that can be reused in different contexts. The constraint block includes its name, such as Newton's Law, the equations, such as force=mass*acceleration, and the definition of the parameters of the equations with their units, which in this example can be mass in kilograms, acceleration in meters per second per second, and force in newtons.

Parametric diagram. The **parametric diagram** in Figure 2.15 shows a constraint called *Orbit Analysis Model* with some of its input and output parameters designated as squares flush with the inside boundary. These parameters include key orbit parameters and selected mission **measures of effectiveness** for *revisit time* and coverage. This *Orbit Analysis Model* in the figure is a proxy for an analytical model that can perform the computation. The corresponding properties of the *Mission Context* and *Spacecraft* are bound to the parameters of the analysis model as indicated by the **binding connectors** with the «equal» key word. The parameter names in the constraint and the property names in the parts do not have to match, although they do in this example, enabling the analytical model to be reused with other design models. The binding relationship enables the integration between the system design model and different analytical models that can execute the analysis. In this way, SysML provides a means to identify critical design properties, reconcile the critical design properties with the parameters of the analytical models, and trace the design properties to requirements.

ARCHITECTING SPACECRAFT WITH SYSML

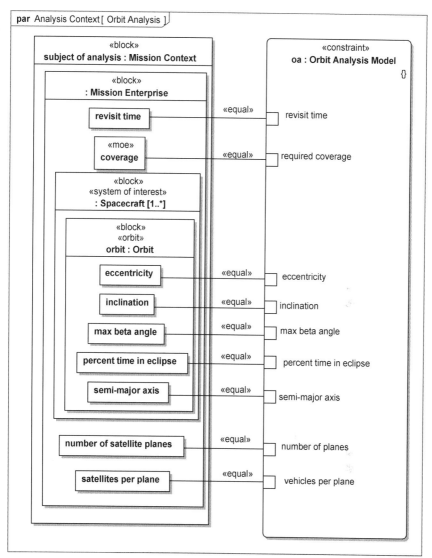

FIGURE 2.15

A parametric diagram binds the parameters of the constraint to the properties of the design.

2.5 Modeling Requirements

Requirement. A text **requirement** can be expressed in SysML and then related to other parts of the model to establish traceability to the system design, analysis, and test cases. This extends traditional requirements management approaches which primarily focus on traceability relationships between requirements. The trace relationships can be navigated to provide robust impact assessments.

Requirement relationships. A SysML model represents a typical document-based specification as a set of requirements which are generally organized into a tree of requirements. A partial tree is indicated in Figure 2.16, where the *Spacecraft Specification* contains the *Functional and Performance* requirement which in turn contains the *Probability of Detection* requirement.

There are several kinds of requirements relationships between requirements and other kinds of elements, that include **satisfy**, **verify**, **refine**, and **derive**. An example of a traceability between a requirement and the corresponding design, analysis, and test case is shown in Figure 2.16. The *Sensing* requirement for the *Payload Sensor* specifies the required *resolution* and *sensitivity*. This requirement is satisfied by the resolution and sensitivity of the *Payload Sensor* as indicated by the satisfy relationship from the *Mid-Range IR Scanner*. The required resolution and sensitivity are derived from the *Spacecraft* system requirements for *probability of detection* and *orbit altitude*. The **rationale** for the derivation is based on the *Sensor Performance Analysis*. The **test case** called *Verify Sensor Resolution* is used to verify the *Payload Sensor* satisfies its *resolution* requirement.

Requirements and their relationships can also be presented in compartments of the requirement and in requirements tables, which provide a compact way to present the requirements and their relationships.

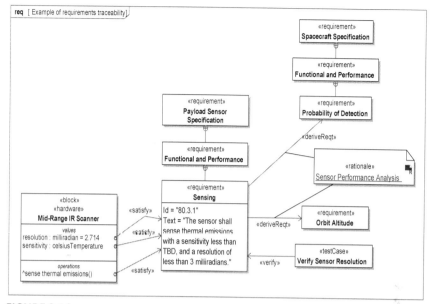

FIGURE 2.16

The Sensing requirement is derived from other requirements with supporting rationale. This requirement is satisfied by the Mid-Range IR Scanner, and is verified by the Verify Sensor Resolution test case.

2.6 Modeling Packages

The SysML model contains model elements. Each model element represents something expressible in SysML, such as a system, a component, a component feature such as its property, function, or interface, or a relationship between components, such as a whole-part relationship. A system model can get very large as details are added. An important aspect of modeling with SysML is the need to manage the model.

This begins with a well-defined **model organization** using **packages** that contain the model elements. A package is like a folder, and provides a way to organize the model into logical groupings. For example, packages can contain the Requirements, Structure, Behavior, and Parametrics as shown in **package diagram** in Figure 2.17. In this example, the *Structure* package contains nested packages for the different levels of design that include *Mission*, *System*, and *Subsystem*. The other packages for *Requirements*, *Behavior*, and *Parametrics* can contain similar nested packages for each level of design. This model organization also includes a *Supporting Elements* package that contains other model elements that cross-cut different levels of design, such as a units library and items that flow through the system.

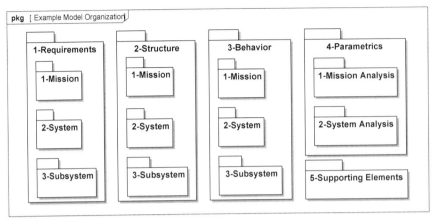

FIGURE 2.17

The package diagram describes the model organization in terms of a set of nested packages.

In this way, a model is organized into a tree of packages, where each package may contain other nested packages and model elements. There is no single way to organize the model no more than there is a single way to organize what goes into your cabinets and drawers. However, a model that is organized based on the product structure and the 4 pillars (Requirements, Structure, Behavior, and Parametrics) is a reasonable place to start.

2.7 Diagram Versus Model

Building the system model consists of populating the packages with model elements that describe the system and environment of interest. This is accomplished by adding elements, modifying elements, and deleting elements from the model using the SysML diagrams as a means for constructing and visualizing the model.

Each diagram is a view that presents a subset of the information contained in the model. For example, an internal block diagram may show the connection between the components and/or subsystems. This diagram is not typically used to present other information contained in the model such as the functions the components perform. Each kind of diagram highlights different aspects of the system.

The model is a tree of model elements, and each model element has a unique position in this **containment tree**. A **fully qualified name** is used to uniquely identify each model element that includes the path from the root element at the top of the containment tree.

A model element may appear on zero, one or many diagrams. For example, the Spacecraft block appears on multiple diagrams, but is defined in only one place in the model containment tree. When a model element is modified, the change to the element is reflected in all diagrams that the element appears. The model serves as a repository of data, where each element is uniquely identified, but can appear in many different views. This model-based approach provides the ability to precisely represent a system, while at the same time provides the ability to ensure consistency among the many different views of a system.

2.8 System Modeling Tool

A system modeling tool is used to build the system model. It provides the basic capabilities to create, modify, delete, and view the model. A modeling tool also provides model checking to ensure the model conforms to the language specification and other user-defined validation rules. A simple example is that a modeling tool should permit a system element to satisfy a requirement, but not allow a requirement to satisfy a system element. There are many other capabilities that different modeling tools offer, such as the ability to execute an activity diagram, the ability to auto-generate documentation, the ability to query the model, and the ability to exchange the data via the tool's application program interface (api). The tool should also support standards that enable integration with other modeling tools and data sources. An example is the ReqIF standard that enables integration with requirements management tools. Each tool has its unique features, and its strengths, weaknesses, and price point. Different vendor tools should be carefully evaluated to select a tool and environment that meets your individual, project, and organizational needs.

A typical **modeling tool user interface** is shown in Figure 2.18. The interface includes the following key elements:

- A browser that depicts the model containment tree
- A diagram area or canvas to create, modify, and view the diagrams
- A pallet with the diagram symbols used to create the diagrams
- A toolbar with the menu options that provide the tool functionality

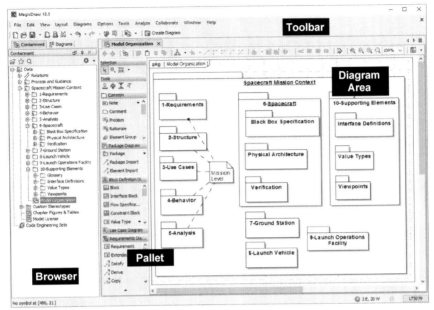

FIGURE 2.18

A typical system modeling tool interface that includes the diagram area, browser, pallet, and tool bar.

The tool typically provides a dialogue box to view the detailed information associated with a particular model element. For example, the dialogue box for a block may include its properties, functions, ports, state behavior, and its relationship to other model elements, as well as a documentation field to provide additional descriptive information. In addition, the modeling tool provides mechanisms to navigate the model, such as locating a model element in the browser. There are many other features available to exploit the rich data set that a system model contains.

MBSE METHOD OVERVIEW

CHAPTER 3

An MBSE method defines how to perform systems engineering activities using a model-based approach. The top level activities for a simplified MBSE method are highlighted in the Mission and System Specification and Design process in Figure 3.1. Although shown in sequence, these activities are often performed concurrently, and can have many interdependencies between them. The activities are consistent with a typical systems engineering process such as those found in the ISO/IEC 15288 standard on System Life Cycle Processes [10]. These activities produce the model artifacts which constitute the system model and are summarized as follows:

- The initial planning activity for the MBSE effort is done as part of the overall project and systems engineering planning. It includes defining the objectives and scope for the modeling effort, tailoring the method, defining the schedule for model artifact delivery, establishing the tool environment, defining the roles and responsibilities to support the modeling effort, and defining the training approach.
- Setting up the model involves defining how the model is organized and establishing the model conventions to support the modeling effort.
- Analyzing the mission and stakeholder needs includes identification of the stakeholders, defining the mission objectives and requirements, and defining the mission context and mission elements that the system of interest interacts with.
- Specifying the system requirements specifies the system as a black box in terms of its functional, interface, performance, physical, and other quality characteristics.
- Synthesizing alternative system architectures defines alternative configurations of system elements, and how the elements interact to satisfy the black box specification.
- Perform analysis is done throughout the development process to evaluate alternative mission and system designs and select a preferred alternative.
- Manage requirements traceability is performed throughout the development process to ensure the mission and stakeholder needs are addressed, and to manage change to the system requirements and design.

- Integrate and verify system is initiated early in the design process to develop a cost effective verification approach to ensure the requirements are verifiable and that the system satisfies its requirements. Early design integration and verification can be performed using simulation and analysis models. Integration and testing of actual hardware and software occurs following the implementation phase.

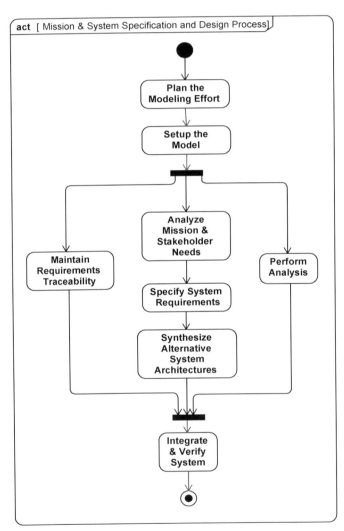

FIGURE 3.1

The Mission and System Specification and Design Process is applied to the design of a spacecraft using a model-based approach.

These activities are applied iteratively to evolve the system model in its breadth, depth, and fidelity throughout the system development lifecycle. The method is applied during the early phases of design to develop the system conceptual design, and then iterated in later phases to evolve the mission and system specification and design through preliminary and detailed design.

The remaining chapters provides an example of how this model-based method can be applied to specify, design, analyze, and verify a complex system such as a spacecraft. Although the activities for this method are described in sequential chapters, in practice they are applied iteratively to meet the objectives for each increment of design and development.

This example leverages the spacecraft design and assumptions from the FireSat II example in Space Mission Engineering: The New SMAD [2]. This example demonstrates the application of a model-based approach with SysML by using model artifacts to capture selected aspects of the mission and system specification and design. A reader of these chapters should refer to the New SMAD for the detailed specification and design, rationale, and assumptions.

More detailed examples of how SysML can be used to support other MBSE methods that include a functional analysis and allocation method and an object-oriented systems engineering method (OOSEM) are included in the modeling examples in Chapters 16 and 17 respectively of A Practical Guide to SysML [8].

CHAPTER 4
PLAN THE MODELING EFFORT

The purpose of this activity is to develop a plan that achieves the modeling objectives within the project constraints. The plan includes defining the overall modeling objectives and scope, identification of key milestones and deliverables, the selection of the MBSE method, tools, and training, the model management process, and the roles and responsibilities for the modeling effort. It is preferred practice to develop a plan for each project phase, such as concept development, preliminary design, and detailed design, that includes identification of design increments and associated deliveries for each phase. For each increment within a phase, the plan identifies the specific increment objectives and intermediate milestones, and the model artifacts to be delivered. As an example, a particular design increment within a preliminary design phase may elaborate selected use cases and perform selected trade studies. The modeling plan should be integrated with the other modeling efforts on the project as part of the overall project planning, and captured in the systems engineering management plan.

A successful modeling plan must adapt to the needs of the project, and at the same time, draw upon the experiences from across the organization and others outside of the organization. The experiences and lessons learned from the Europa Project [11, 12] provide one such example.

This chapter addresses the following:

- Define the modeling objectives and scope to support the project
- Pre-populate the model with existing data about the system and its environment
- Establish the schedule to develop the modeling artifacts
- Select the MBSE method and tools
- Determine how the modeling effort will be organized
- The training needs to support the modeling effort

4.1 Modeling Objectives and Scope

The purpose of the modeling effort for each project phase and design increment must be clearly defined and reflected in clear modeling objectives. The objectives for this spacecraft model are to ensure a consistent and cohesive description of the spacecraft architecture that spans multiple discipline and subsystem views (e.g., Power Subsystem, Avionics Subsystem, Mass Properties, Reliability and Fault Management), and to ensure the quality and flow down of the mission, system, subsystem, and component requirements to satisfy the stakeholder needs. The modeling objectives for a particular design increment are to create or update selected model artifacts such as the mass equipment list, system block diagram, interface specifications, and support the analysis and trade studies to satisfy the critical information needs for the increment.

The scope of the effort defines the breadth, depth, and fidelity of the model to meet its intended purpose. For example, during the early phase of design, the model represents the system architecture at a level needed to define mission scenarios, identify system design alternatives, define the preliminary system breakdown, and support preliminary mass and power estimates, but does not require the model to include the detailed component interfaces and software requirements that may be required for later phases of development.

4.2 Pre-populate the Model

The model-based effort does not generally start from scratch. There is often considerable requirements, design, verification, and analysis data to leverage for the modeling effort. This data is often in many different formats such as excel tables and word documents, and others. The data must be validated and properly transformed prior to including it in the model. The planning should include identification of the data and validation process as well as when the data is needed to support the overall modeling effort. The validated data is then incorporated into the model to support the need.

An organization with mature modeling practices may provide access to enterprise asset repository that include reference models and component reuse libraries. These assets can serve as a starting point for the spacecraft design model. The models developed by each project can be leveraged to further evolve the modeling assets in the enterprise asset repository.

4.3 Schedule the Modeling Artifacts

The schedule for generating the modeling artifacts should support the overall project schedule and internal and external deliverable requirements. The modeling plan identifies which model artifacts are needed to support each project milestone, and the associated maturity of those artifacts at each milestone. The table in Figure 4.1 shows an example at a summary level indicating the level of maturity from Low to Medium to High that is required to support each milestone. The increment plans can define intermediate milestones from one milestone to another to develop the more specific model artifacts for requirements, structure, behavior, and parametrics. The increment schedules should include peer reviews of the model artifacts as part of the technical baseline management process.

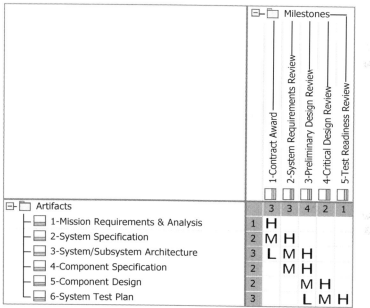

FIGURE 4.1

The matrix shows how the modeling artifacts mature from Low to Medium to High as the project progresses through different phases of development.

4.4 Establish the MBSE Method and Tool Environment

The MBSE method and applicable modeling practices are selected to achieve the model purpose and scope. The method may implement various aspects of the projects organization's systems engineering process such as the method described in this book. The specific modeling practices should be piloted and validated to ensure they yield the desired results. For example, the practice for modeling failure modes should facilitate the failure modes and effects analysis (FMEA).

The model management approach is established to manage updates to the model. During the early phase of design, the control on the baseline is less constrained than later phases. Typically, the configuration management process enables different users to check out parts of the model, make changes, and check the model back in.

The modeling tools are often selected based on organizational considerations, since the organization makes substantial investments in maintaining its engineering tools and environments. However, there are often specific customizations and integrations with other tools that are needed to support the project specific model purpose, scope, and practices.

4.5 Organize the Modeling Effort

The organizational roles and responsibilities for the people involved in the development and use of the model evolve across the project life cycle. During the early phases of a project, a small core modeling team defines the modeling guidelines and practices, and performs the modeling to ensure the modeling practices are adhered to. The broader team members provide input to the core team and review the model artifacts to ensure their inputs are accurately reflected in the model. As the modeling practice matures on the project, custom interfaces such as web browsers may be developed to enable the broader team to enter data directly into the model, and view data from the model in ways that are meaningful to them. The core modeling team continues to ensure the overall integrity of the model, and adapts the modeling practices and tools to support the different needs of the project.

4.6 Establish the MBSE Training Approach.

The training needs for the modeling effort include who needs what training, when the training is needed, and how it will be delivered. For example, in the early phases of a project, the core modeling team requires more intensive training in the modeling language, tool, method, and practices, whereas other members of the project team may only require training to enable them to interpret the model artifacts.

SETUP THE MODEL

CHAPTER 5

This purpose of this activity is to prepare the model by establishing the initial model organization and identifying the modeling conventions and standards to be used by the project.

This chapter addresses the following:
- Organize the Model
- Establish modeling conventions

5.1 Organize the Model

This purpose of this activity is to define the model package structure to organize the mission and system specification and design data. A well-defined model organization facilitates navigation of the model, access control, and reuse. A model organization is analogous to a document structure where the mission and system specification and design documentation are often organized in a folder structure of a document repository. The system model is organized into a package structure as described in Chapter 2.6.

The model organization in Chapter 2.6 is somewhat simplified to illustrate the concepts for organizing a model. A more robust model organization for the spacecraft model is shown in the package diagram in Figure 5.1 The spacecraft model consists of a top level package with nested packages that contain model elements. The model elements in different packages can be related to one another.

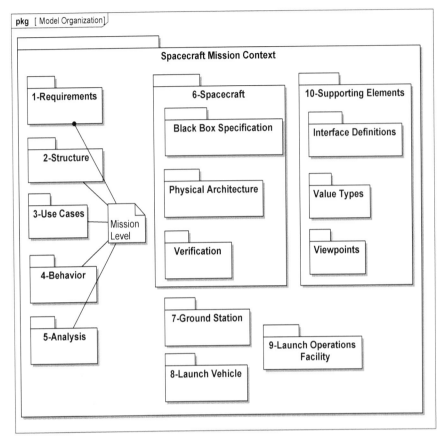

FIGURE 5.1

The Model Organization includes packages to capture specification and design data for the mission, Spacecraft, other mission elements, and supporting model elements.

The top level package is called the Spacecraft Mission Context, which contains nested packages that define the mission specification and design including the mission Requirements, Structure, Use cases, Behavior, and Analysis. The Spacecraft package contains nested packages to define the spacecraft specification and design including the Black Box Specification, Physical Architecture, and Verification. The Ground Station, Launch Vehicle, and Launch Operations Facility packages contain model elements that describe them.

The Supporting Elements package contains supporting information in the Interface Definitions, Value Types, and Viewpoints packages. The Interface Definitions package contain model elements that can be reused throughout the system, such as flows, signals, and port definitions. The Value Types package contains a library of quantities and units used in the model. The Viewpoints package contain the viewpoints that identify stakeholders and their concerns.

The Browser view of the model organization is shown in Figure 5.2 showing some of the nested packages contained within the Spacecraft Physical Architecture package.

These packages are initially empty containers, but are populated with data as one applies the MBSE method. This iterative process enables continuous refinement of the information about the mission and system specification and design. The remainder of this section describes the model elements that are contained in these packages.

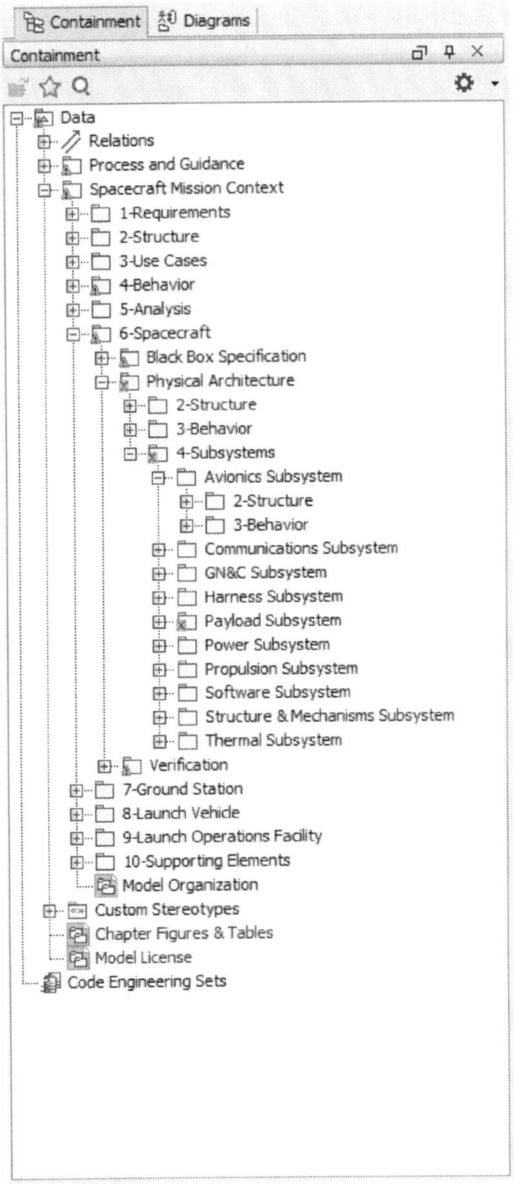

FIGURE 5.2

A Browser View of the Model Organization shows the nested elements contained within the physical architecture package and additional top level packages.

5.2   Establish Modeling Conventions

The Model Setup activity also includes establishing modeling conventions to be applied throughout the modeling effort. Examples including naming conventions for different kinds of elements, diagram naming conventions, diagram layout conventions. , and selection of what kinds of diagrams are used to represent different aspects of the system (e.g., activity diagram versus sequence diagram). For this example, some naming conventions include:

- Blocks, Activities, and other Classifiers begin with upper case
- Parts, properties, and actions begin with lower case
- Port names are appended with i/f (for interface)
- Activity names are defined as verb-noun

Other modeling conventions provide guidance for annotating the model which can be an important part of the model content, and can be used to support the generation of documentation from the model. For example, model elements can include a text description that can be included in a Glossary that is part of the model which helps to define a consistent vocabulary for the project, Annotations are also used to capture design rationale. Other examples of modeling conventions define how to specify different kinds of interfaces such as software, electrical, and mechanical interfaces.

ANALYZE MISSION & STAKEHOLDER NEEDS

CHAPTER 6

The purpose for this activity is to analyze the mission and stakeholder needs and establish the mission objectives, measures of effectiveness, mission requirements, and mission concept to address these needs. This activity includes identifying the stakeholders and determining what problem needs to be addressed by the mission.

For this FireSat II mission, the problem to be addressed is the damage caused by wildfires across the United States and Canada, and the need is to provide data that can help mitigate this damage. The mission objectives, measures of effectiveness, and mission requirements are derived from the needs. The mission concept includes the definition of the mission elements and their required functions and performance to satisfy the mission requirements within the specified technical, cost, schedule, regulatory constraints, and risk thresholds.

This activity can also support defining concepts that address a range of missions, such as remote sensing, that could be addressed by a class of spacecraft, but the focus for this example is on the specific FireSat II mission.

This chapter addresses the following:

- Identify the stakeholders and the problems to be addressed.
- Define the mission objectives.
- Develop the mission requirements that support the mission objectives.
- Identify the external mission elements that interact with the system of interest (i.e., the Spacecraft)
- Define the measures of effectiveness (*MoE's*) that can be used to quantify the mission and stakeholder value of a proposed solution.
- Perform mission analysis.
- Define the mission level behavior including the mission functions and event time-line.
- Identify potential mission failure modes.

6.1 Identify Stakeholders and Their Concerns

A problem is a deficiency of the current state of an enterprise of interest from a stakeholder perspective. There are many techniques used to analyze the current state, perform causal analysis, and identify problems. Capturing the current state of the system and enterprise in a model can facilitate problem identification. The fundamental problem to be addressed for the FireSat II mission is the damage and associated costs resulting from forest fires across the United States and Canada, and the need for the FireSat II mission is to provide data that can help mitigate this damage.

There are often many stakeholders with different interests in the mission. The identification of stakeholders continues throughout a project. Some of the FireSat II stakeholders are shown in Figure 6.1. For this example, the primary stakeholders include the end-users, which are the Forest Service and Fire Department, the Mission Sponsor, the Development Contractor, and the spacecraft Operator. The **viewpoints** shown in the figure are used to capture stakeholders and their concerns in terms of what they care about, and identify the relevant information from the model to address their concerns.

A **view** conforms to a particular viewpoint by presenting the relevant information from the system model to address the stakeholder concerns. A view can be presented as a diagram, a table, or an entire document. The format of the artifact can also be specified, such as a document in word, pdf, or html. The documents, such as specifications, interface control documents, and architecture description documents, can be generated from the model by querying the system model, and rendering the information the specified document format.

Viewpoint and view are important architecture concepts which provide a means to explicitly capture and address a broad spectrum of stakeholder concerns in the model. The viewpoint and view concept provide a mechanism to query the model for the information that addresses these concerns, and present the resulting information in a way that is useful to each stakeholder.

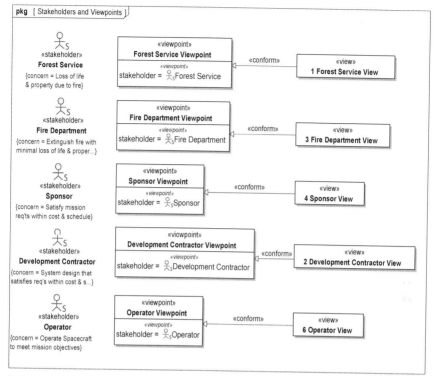

FIGURE 6.1

The viewpoints capture the stakeholders and their concerns, and the conforming views present the specific information from the model to addresses their concerns.

For FireSat II, the Forest Service is concerned about loss of life and property due to forest fires. The Fire Department is concerned about their ability to put out a fire in a timely manner to prevent the loss of life and property. The Sponsor cares about the ability to satisfy the mission requirements within cost and schedule constraints. The Development Contractor is concerned about establishing a feasible design to satisfy the mission requirements, and the Operator is concerned about his or her ability to operate the spacecraft to achieve the mission objectives. An example of a view is the Operator View that includes the command and telemetry definition needed to support mission operations. As the project evolves, other stakeholders are identified and defined in the model in a similar way. Each set of stakeholder concerns can impose additional requirements and constraints on the mission and system.

The stakeholders, viewpoints, and views are captured in the Viewpoints package shown in the model organization in Figure 5.1.

6.2 Define the Mission Objectives and Requirements

The mission objectives should reflect stakeholder value and address their concerns. For the New SMAD, a primary objective of the Forest Service is for FireSat II to 'Detect and monitor forest fires in US and Canada using as little financial resources as possible', and an Operator objective is to 'Provide forest fire data to the end user in real time, archive data, monitor and maintain the health and safety of FireSat II.' These objectives are reflected in the use case diagram in Figure 6.2. As emphasized in The New SMAD, the mission and system specification and design process is intended to provide balanced solutions that meet these objectives within cost, schedule, and risk constraints.

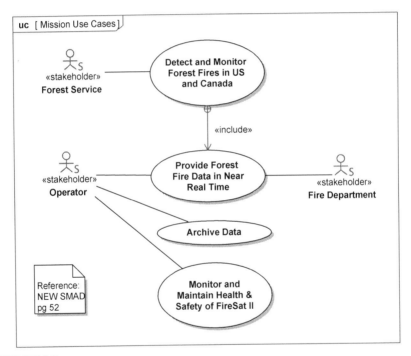

FIGURE 6.2
The use case diagram reflects the mission objectives.

The mission requirements are specified to support the mission objectives. The New SMAD specifies the FireSat II mission requirements that are shown in the table in Figure 6.3. These mission requirements are the source mission requirements specified by the sponsor, which in this case is the contracting organization. In practice, the sponsor often provides mission requirements to the spacecraft developer. These requirements are captured in the model, and can be presented in both tabular and graphical views.

#	Id	Name	Text
1	34	Mission Requirements-SMAD Table 3-4	
2	34.1	Performance	
3	34.1.1	Weather	Work through light clouds
4	34.1.2	Resolution	50 meter resolution
5	34.1.3	Geo-location Accuracy	1 km geolocation accuracy
6	34.2	Coverage	Coverage of specified forest areas within the US at least twice daily.
7	34.3	Interpretation	Identify an emerging forest fire within 8 hours with less than 10% false positives
8	34.4	Timeliness	Interpreted data to end user within 5 minutes
9	34.5	Secondary Missions	Monitor changes in mean forest temperature to +/- 2 C
10	34.6	Commanding	Commandable within 3 min of events. download units of stored coverage areas.
11	34.7	Mission Design Life	8 years
12	34.8	System Availability	95% excluding weather. 24 hour maximum downtime
13	34.9	Survivability	Natural environment only. not in radiation belts.
14	34.10	Data Distribution	Up to 500 fire-monitoring offices + 2,000 rangers worldwide (max c 100 simultaneous users)
15	34.11	Data Content, Form, and Format	Location amd extent in lat/long for local plotting. avg temp for each 40 m2
16	34.12	User Equipment	10 X 20 cm display with zoom and touch controls, built-in GPS quali map
17	34.13	Cost	Non-recurring<$10M Recurring<$3M/year
18	34.14	Schedule	Operational within 3 years
19	34.15	Risk	Probability of success>90%
20	34.16	Regulations	Orbital debris, civil program regulations
21	34.17	Political	Responsive to public demand for action
22	34.18	Environment	Natural
23	34.19	Interfaces	Interoperable through NOAA ground stations
24	34.20	Development Constraints	None

FIGURE 6.3

The source mission requirements specified by the sponsor are captured in the model as defined in Table 3-4 of the New SMAD.

The New SMAD source requirements are subject to further clarification of intent by the spacecraft development contractor. As a result, the development contactor further refines the source mission requirements based on analysis of the mission use cases and scenarios. An example of the refined mission requirements and their relationship to the source mission requirements is shown in Figure 6.4. Mission performance analysis is used to derive requirements from the refined mission requirements. This is discussed in Section 6.4.

A fundamental mission requirement is to Detect and Monitor Forest Fires in US and Canada. This requirement traces directly to the primary mission objective shown in the use case diagram in Figure 6.2, and is further elaborated by other mission requirements that specify the Coverage,

Response Time, Probability of Fire Detection, False Alarm Rate, Location Accuracy, and Weather Environment.

Additional mission requirements relate to Launch Vehicle Compatibility, Ground Station Compatibility, and Regulations such as the constraints on the available Radio Frequency Spectrum. The mission requirements also specify Data Distribution and Delivery, Data Archive, Mission Availability, Spacecraft Survivability, Mission Life, Lifecycle Cost, and Launch Schedule.

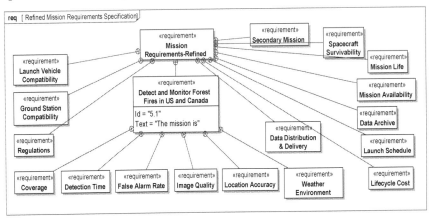

FIGURE 6.4
The Mission Requirements-Refined are refined by the spacecraft development contactor from the source mission requirements shown in the previous figure.

Clearly, there is considerable analysis needed to derive the mission and system requirements from the set of stakeholder concerns and needs. The mission analysis provides the supporting rationale for the requirements. The orbit analysis is a key mission analysis, and is discussed later in this chapter.

The matrix view in Figure 6.5 shows the refinement relationship between the refined mission requirements in Figure 6.4 and the New SMAD mission requirements in Figure 6.3. The arrows in the cells of the matrix indicate that the mission requirements in the columns refine the source requirements in the rows.

6.3 Define the Mission Context

The top level block definition diagram in Figure 6.6 sets the context for the Spacecraft, which is the system of interest in the diagram and focus of this design effort. The context is defined by identifying the mission elements that the Spacecraft directly or indirectly interacts with throughout its mission that includes other systems, users, and its external environment.

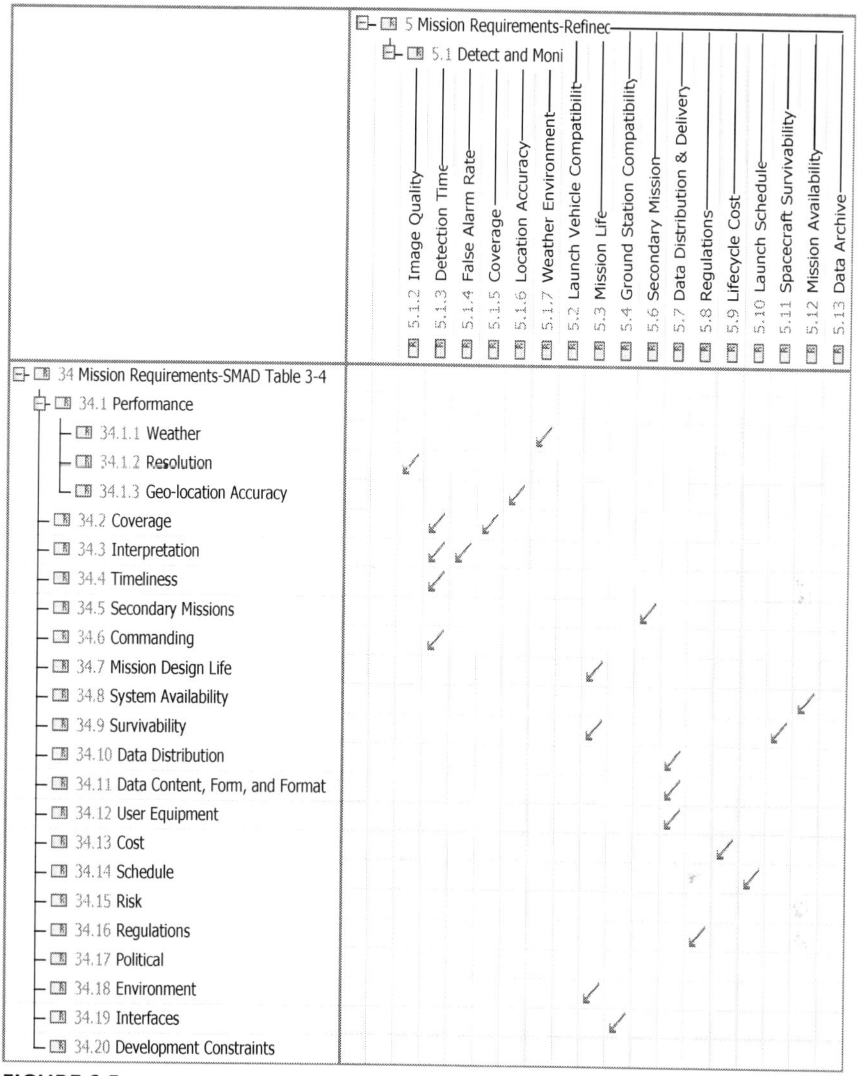

FIGURE 6.5

Traceability between the refined mission requirements specified in Figure 6.4 and the source mission requirements from the New SMAD in Figure 6.3. The line with the arrowhead depicts a refinement relationship indicating that the mission requirements in the columns refine the source mission requirements in the rows.

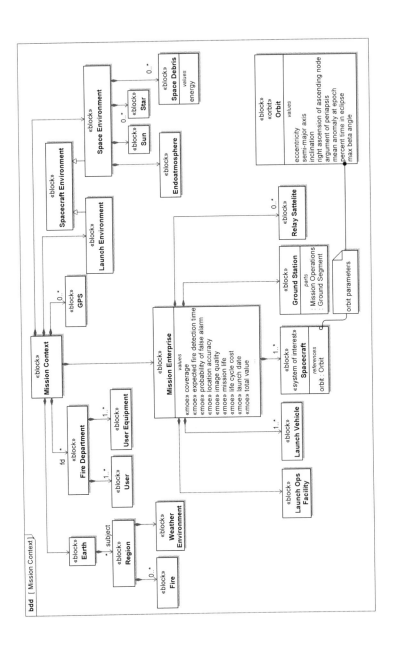

FIGURE 6.6

This block definition diagram identifies the mission elements that the Spacecraft directly or indirectly interacts with throughout its mission.

The top level Mission Context block aggregates the other mission elements that includes the Mission Enterprise. The Mission Enterprise further aggregates the primary mission elements needed to accomplish the mission that includes the Spacecraft, Launch Vehicle, Launch Ops Facility, Ground Station, and Relay Spacecraft. The Ground Station is composed of Mission Operations and the Ground Segment as shown in its parts compartment. Also note that the Mission Enterprise is composed of 1 to many Spacecraft, 1 to many Launch Vehicles, and 0 to many Relay Satellites as indicated by the multiplicity on the ends of the whole-part relationships with the black diamond.

The Mission Enterprise is actually composed of 1 to many Launch Systems, where each Launch System is composed of a single Launch Vehicle and 1 to many Spacecraft. The Launch System is not represented to simplify this model. The number of Launch Systems and the number of Spacecraft per Launch Vehicle is the subject of a trade study.

The Mission Context also includes the Spacecraft Environment, which is further classified into a Launch Environment and Space Environment. The Space Environment is composed of several planetary objects that include the Sun and Stars that the Spacecraft interacts with. This includes the solar radiation affects from the Sun and star light energy that the Spacecraft receives and uses for navigation. The Space Environment also includes the Endoatmosphere, which induces drag on the Spacecraft, and Space Debris that can impact the Spacecraft. The Fire Departments are also shown in this figure, along with the regions on the Earth that are intended to be monitored by the Spacecraft. The Regions that are covered are referred to as the subject, and these Regions can include Fires.

The overall critical mission performance parameters are called measures of effectiveness (MoE). These parameters are used to evaluate and optimize the mission effectiveness for each design alternative. The MoE's are captured as value properties of the Mission Enterprise and designated with the keyword «moe». The mission performance requirements such as Coverage and Detection Time that are specified in Figure 6.4, are traced to the corresponding MoE's of the Mission Enterprise.

6.4 Perform Mission Analysis

The mission analysis is performed as part of the Perform Analysis activity shown in Figure 3.1 and described in Chapter 9. This analysis is performed concurrently with the Analyze Mission and Stakeholder Needs activity to derive critical mission and system performance requirements needed to optimize the mission measures of effectiveness identified in the previous section. In particular, the orbit analysis is used to establish the Spacecraft orbit, which is used to derive many other system requirements.

The block called Orbit in Figure 6.6 contains the key properties needed to specify an orbit. The Spacecraft contains an orbit property in its block compartment that references the Orbit block as shown in the figure. The mission and system analysis is performed to select the orbit parameters and other system design properties to maximize mission and stakeholder value based on its MoE's.

The analysis model used to specify the orbit is highlighted in Figure 6.7. In this example, the orbits are analyzed using a COTS simulation tool from AGI called Systems Tool Kit (STK). The figure depicts four Spacecraft in two orbital planes. The orbital planes differ in the right ascension of the ascending node (RAAN), and the two Spacecraft in each plane are separated by 180 degrees. This constellation configuration is the proposed alternative that provides the necessary coverage and revisit times based on the analysis described below.

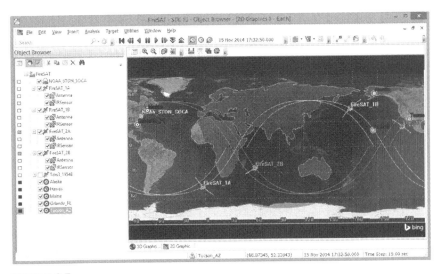

FIGURE 6.7

Orbits for the 4 Spacecraft Constellation with 2 Spacecraft in each of 2 planes provides necessary coverage and revisit times.

The analysis results summarized in Figure 6.8 are used to evaluate and select the orbit parameters and other key mission parameters such as the number of spacecraft. Chapter 9 describes how the SysML parametric diagram is used to specify the input and output parameters of the analysis, such as the orbit analysis model described here.

This study investigated the impact of the orbital altitude and number of Spacecraft on the revisit times for representative locations in the required coverage regions. The mission requirement is to provide coverage of specified forest areas within the US and Canada at least twice daily, although this particular analysis addressed the complete US only. A single Spacecraft at 500 kilometer altitude does not provide the needed coverage. Placing the Spacecraft in a higher orbit at 700 kilometers increases the total coverage and decreases revisit times as shown in the figure. This study shows that a single Spacecraft at 700 kilometers is not capable of achieving the desired revisit times, and adding a second Spacecraft in the same orbital plane only marginally improves this performance (refer to 2 Sat data). However, adding a second plane that phases the orbits with two additional Spacecraft decreases revisit times nearly in half (refer to 4 Sat data). Adding a third plane further decreases revisit time, but with diminishing returns for the additional cost.

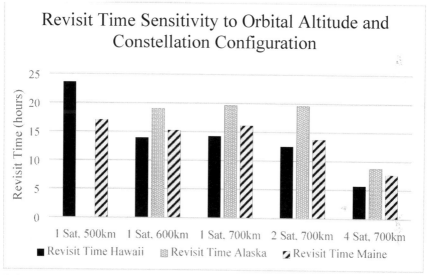

FIGURE 6.8

Orbit analysis results showing the Revisit Time to selected regions within the specified coverage area.

The 4 spacecraft configuration is the proposed mission design alternative. Since there are 2 spacecraft in each plane, it is assumed that two Launch Vehicles each carry 2 Spacecraft to the desired orbit plane. However, this is subject to further trade study and analysis to balance with other system design considerations related to size, payload sensor, communications, and propulsion.

The selected orbit is sun synchronous with an orbit period of 96 minutes. The above Orbit Analysis provides the rationale to derive the requirements for Orbit Altitude, Orbit Plane, and the Number of Spacecraft from the mission requirements for Coverage and Detection Time as indicated in Figure 6.9. The Revisit Time is derived from the Detection Time. The Number of Launch Vehicles is further derived from the Number of Spacecraft and the Orbit requirements. As noted previously, the orbit requirements must be balanced with other system requirements to ensure the Coverage, Detection Time, and other mission requirements are satisfied.

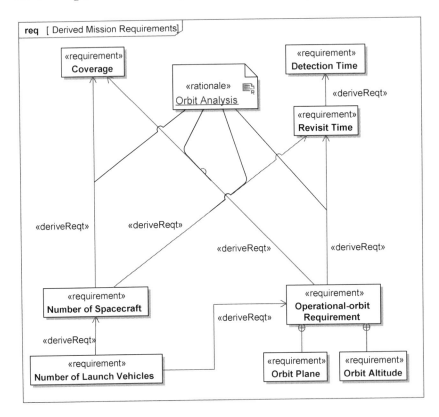

FIGURE 6.9

Traceability showing the requirements for Orbit Plane, Orbit Altitude, Number of Spacecraft, and Number of Launch Vehicles derived from mission requirements.

6.5 Specify Mission Behavior

The mission behavior includes specifying the mission functions, event timeline, and identification of potential mission failures. The top level mission functions needed to Perform Mission are shown in Figure 6.10. These include the functions to Launch Spacecraft, Separate from Launch Vehicle, Control Trajectory, Deploy Mechanisms, Maintain Spacecraft Operations, and Provide Observation Data. Each of these mission functions may involve interactions throughout the mission between the Spacecraft and other mission elements such as the Launch Vehicle, Ground Station, and the Space Environment.

Mission analysis is used to derive the system performance and physical requirements associated with each function. For example, the Control Trajectory function must satisfy the orbit requirements that are derived from the Orbit Analysis described in the previous section.

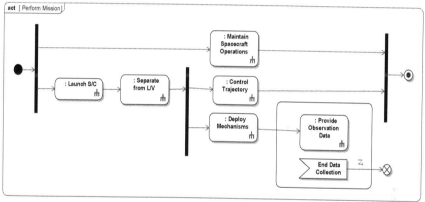

FIGURE 6.10

Top level functions of the Perform Mission activity.

The mission functions are further decomposed as part of Specify System Requirements activity described in the next chapter to specify what the spacecraft must do to support the mission functions. The Maintain Spacecraft Operations activity diagram in Figure 6.11 is an example that shows that the Spacecraft is required to Provide Telemetry Data, Receive Ground Commands, Provide Electrical Power, Control Thermal Environment, and Manage Faults.

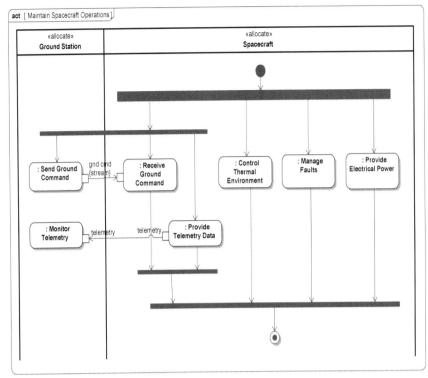

FIGURE 6.11

Some of the functions the Spacecraft must perform are shown in the Maintain Spacecraft Operations activity.

The mission timeline specifies the sequence of mission events as shown in the sequence diagram in Figure 6.12. The events are associated with the sending of signals between different mission elements. The Power On signal from Launch Ops Facility to the Spacecraft results in a state change of the Spacecraft from the off state to the on state. Other states are not shown for brevity. The more detailed sequence diagram for On Orbit & Data Distribution Events is referenced in Figure 6.12 and shown in more detail in Figure 6.13. Time constraints can be specified between any two events, as indicated by the time duration between the time a fire starts and the time the data is made available to the end user. The sequence diagram in Figure 6.13 also includes some more advanced control features to represent loops and parallel sequences (i.e., par).

ARCHITECTING SPACECRAFT WITH SYSML

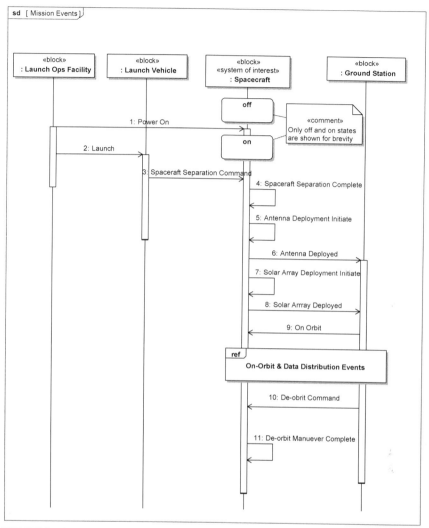

FIGURE 6.12

Sequence diagram showing the sequence of Mission Events including a reference to a more detailed sequence diagram for the On-Orbit & Data Distribution Events.

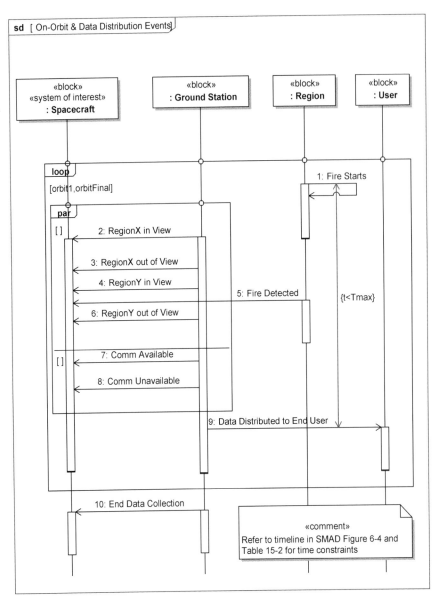

FIGURE 6.13

Sequence diagram showing the sequence of the On-Orbit & Data Distribution Events. This sequence diagram is referenced in the sequence diagram in Figure 30.

It is important to identify the top level failure modes that can cause a mission failure, and design the mission and system to mitigate these failures. Figure 6.14 shows potential failure modes for each top level function described in the Perform Mission activity in Figure 6.10. There are generally multiple failure modes associated with failing to perform a particular activity, although for simplicity, only a single abstract failure mode for each activity is shown in this figure.

Each failure mode in the figure is shown to violate an activity. A failure mode is a constraint violation for a particular activity where one or more performance or physical properties are outside of their acceptable range for some period of time. The violation of one constraint can be caused by another constraint violation. The simple cause effect relationship may result from very complex underlying phenomena that may be difficult to characterize and quantify. The elaboration of these failure modes are illustrated in Figure 8.19, and are intended to aid in the failure analysis and failure mitigation strategy.

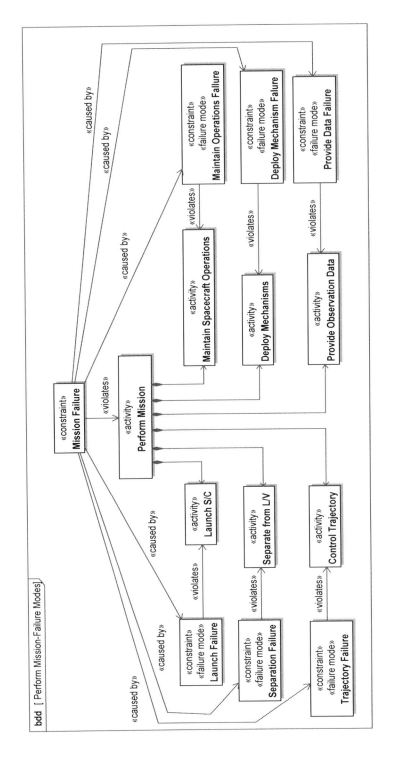

FIGURE 6.14

The potential mission failure modes are identified from the Perform Mission activity. The potential failure modes are defined in terms of violations to satisfy the constraints associated with each activity. A failure mode can be

SPECIFY SPACECRAFT REQUIREMENTS

CHAPTER 7

This purpose for this activity is to specify the black box requirements for the Spacecraft that support the mission requirements. A black box requirement specifies functional, performance, physical, and interface requirements that are observable and verifiable at the Spacecraft level. The mass requirement for the Spacecraft is an example of a requirement that can be measured by observing the Spacecraft as a black box, whereas the mass of a Star Tracker is internal to the Spacecraft and not measurable when observing the Spacecraft as a black box. Similarly, the requirement to return observation data is observable as an output of the Spacecraft, whereas the requirement to process command data yields outputs that are not externally observable. Some requirements associated with managing the Spacecraft internal state, such as provide electrical power, are treated as black box requirements, even though the internal power is not externally observable. Specifying the black box requirements at the right level of abstraction avoids over-constraining the Spacecraft design that can enable broader exploration of the Spacecraft design alternatives to satisfy the requirements. This in turn can yield a more cost-effective design solution.

Several different modeling artifacts are used to specify the spacecraft external interfaces with the other mission elements, the spacecraft functions that it must perform to support the mission functions, the spacecraft states, and the spacecraft critical performance, physical, and quality characteristics.

This chapter addresses the following:

- Define the Spacecraft external interfaces with other mission elements.
- Specify spacecraft behavior to support including the system functions and states.
- Identify potential system failure modes.
- Specify the spacecraft performance, physical and other quality characteristics
- Capture the spacecraft system requirements that support the mission requirements as a black box specification that is used to refine the text-based requirements specification

7.1 Define Spacecraft External Interfaces

The internal block diagram for the Mission Context in Figure 7.1 shows the Spacecraft external interfaces with the other mission elements that are identified in Figure 6.6. The Spacecraft, Launch Vehicle, Launch Ops Facility, and Ground Station are all parts of the Mission Enterprise. The ports on the Spacecraft boundary identify each of its external interfaces. The item flows depicted as black filled arrows on the connectors, represent flows between the mission elements and the Spacecraft.

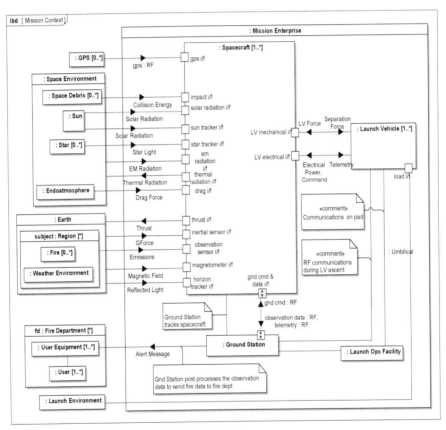

FIGURE 7.1

Spacecraft external interfaces with other mission elements are identified on the Mission Context internal block diagram.

The ports represent diverse kinds of interfaces including electrical, mechanical, radio frequency (RF), and other physical interfaces. More generally, they can represent software interfaces, user interfaces, and more abstract logical interfaces. The gnd cmd & data i/f port is the interface that specifies the external interface to the Ground Station. As indicated in the figure, the item flows include the ground commands from the Ground Station to the Spacecraft, and the observation data and telemetry from the Spacecraft to the Ground Station. These flows are all encoded as RF signals. This particular port is later seen to be delegated to the interface for the Spacecraft antenna.

Other ports include the external electrical and mechanical interfaces to the Launch Vehicle called LV electrical i/f and LV mechanical i/f. While on the pad, the commands and telemetry are sent to and from the Spacecraft through the LV electrical i/f to the Launch Vehicle and from the Launch Vehicle to the Launch Ops Facility through an electrical umbilical. During Launch Vehicle ascent, the Spacecraft continues to communicate through the Launch Vehicle via RF communications with the Launch Ops Facility, which in turn communicates with Ground Station. Once the Spacecraft separates from the Launch Vehicle and its antenna is deployed, the Spacecraft communicates with the Ground directly through its gnd cmd & data i/f port. The interaction between the different mission elements is reflected in the mission events sequence diagram in Figure 6.12.

The port called observation sensor i/f accepts the Emissions from a subject:Region on the Earth, and specifies the interface to the Spacecraft Payload Sensor. There is also an Alert Message from the Ground Station to the Fire Department that is encoded in a web-based protocol such as HTTP.

The solar radiation i/f specifies an external Spacecraft interface to collect the solar radiation from the Sun, which is later seen to provide the interface for the Solar Array. The port called the impact i/f is a very abstract interface that corresponds to the exposed surface of the Spacecraft that is subject to impact with Space Debris. Different filtered views of this Mission Context can highlight the ports and connectors that are used during each mission phase (e.g., pre-Launch, Ascent, on-Orbit).

Note that only some of the mission elements have ports. It is up to the modeler to determine which ports are needed to specify the interface. For example, it does not necessarily make sense to specify a port on the Sun, unless we want to specify more detailed information regarding the solar radiation at the Sun's surface.

7.3 Specify Spacecraft Behavior Requirements

The Spacecraft black box functions define what the Spacecraft must do to support the mission functions. These functions are derived by further elaborating each mission function (i.e. activity) in the Perform Mission activity diagram in Figure 6.10. As described in the introduction to this section, the intent is to specify a function the system performs without over-constraining how the function is performed by its components. For example, the functions in Figure 6.11 to Provide Telemetry, Provide Electrical Power, Control Thermal Environment, and Manage Faults are black box system functions that can be realized by alternative Spacecraft system design concepts.

Each black box function has associated interface and performance requirements associated with it. For example, the Spacecraft function to Provide Telemetry requires an external interface between the Spacecraft and Ground Station. The performance requirement for the Spacecraft function to Provide Electrical Power specifies the power levels that the Spacecraft must provide throughout the mission. The performance and physical requirements are derived based on further engineering analysis as described later in Section 7.4.

Another example of Spacecraft black box functions are the Control Acceleration and Control Attitude functions that are derived from Control Trajectory in the Perform Mission activity. The associated performance requirements for these functions include the attitude control accuracy and acceleration levels.

The overall Spacecraft behavior is described in the state machine diagram in Figure 7.2. This state machine specifies the Spacecraft states and the events that trigger transitions from one state to another during different mission phases. An example is the receipt of a Power On signal sent from the Launch Ops Facility to the Spacecraft during the pre-launch phase that was shown in the sequence diagram in Figure 6.12 that resulted in a transition from the off state to the on state. The activities that are performed by the Spacecraft in each state are also specified, such as the Spacecraft functions in the on state which support the Maintain Spacecraft Operations activity.

The on state contains two concurrent regions which are separated by a dashed line. The region in the lower part of the spacecraft specifies whether the Spacecraft is in normal or safe state. The other region specifies whether the Spacecraft is in the launching or operating state. The operating state contains nested states that include separating, transferring, orbiting, and de-orbiting. At any given time, the Spacecraft is in exactly one state in each region.

The transition from one state to another may include a guard condition that must be true for the transition to occur, such as the transition from the not sensing state to the sensing state that is triggered by receipt of the Sensor On Command and subject to the condition that Region in View is true.

ARCHITECTING SPACECRAFT WITH SYSML

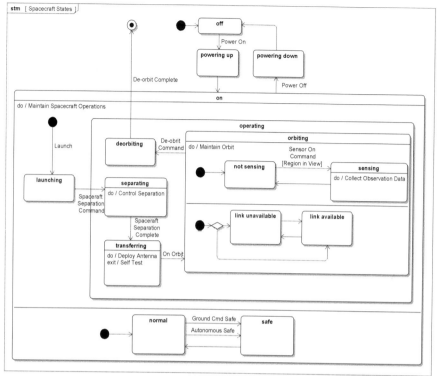

FIGURE 7.2

The Spacecraft state machine diagram specifies the overall behavior of the Spacecraft in terms of its states, transitions between states, and behaviors within each state.

The basis for identifying mission failure modes was discussed in the previous chapter and shown in Figure 6.14. The Spacecraft and other mission elements, such as the Launch Vehicle, can each contribute to a Mission Failure. The Spacecraft system failure modes are identified in a similar way as the mission failure modes. The failure modes reflect the potential inability of the Spacecraft to perform each of its functions within its specified performance bounds. There are generally multiple failure modes for each system function.

As noted earlier, the Spacecraft functions to Control Attitude and Control Acceleration contribute to the mission function to Control Trajectory. The inability to maintain the attitude rate and/or maintain the steady state attitude within the specified performance bounds are two potential failure modes for Control Attitude. The discussion on component failure modes in Chapter 8.3 provides an example of how lower level failure modes can contribute to a mission failure.

7.4 Specify Spacecraft Performance

The Spacecraft black box performance and physical properties and other quality characteristics reflect desired characteristics of the Spacecraft. As an example, the mass property reflects the required mass of the Spacecraft, the delta-v property reflects how much Delta-V (velocity change) the Spacecraft must provide to meet mission requirements, the power property reflects how much electrical power the Spacecraft must provide, and the reliability property reflects the specified MTBF of the Spacecraft. The required values are subject to trade study to achieve the best overall mission cost-effectiveness. The best estimate of these property values are estimated based on analysis of the Spacecraft design and compared with the required value. This estimate will vary as the design matures. Early in the design process, a larger reserve is included to account for design uncertainty.

The system analysis is performed as part of the Perform Analysis activity shown in Figure 3.1 and described in Chapter 9. Similar to the mission analysis, this analysis is performed concurrently with the Specify System Requirements activity to specify the critical system black box performance, physical, and other quality characteristics. The required property values and the estimated design values are updated based on iterative analysis of the mission design and the system design.

As an example, the Spacecraft mass is constrained by the Launch Vehicle and other factors. The estimated mass is derived from the initial estimates of the dry mass of the Spacecraft plus the propellant mass. The initial estimate of the dry mass is often based on general rules of thumb and comparisons with similar designs. The propulsion analysis uses the dry mass estimate to compute the propellant mass needed achieve and maintain the specified mission orbit. This includes the delta-V required to transfer from launch vehicle separation to operational orbit, maintain the orbit, perform the de-orbit, and perform other required Spacecraft maneuvers, such as the "yaw flip" to ensure the solar array maintains exposure to the Sun.

The table in Figure 7.3 shows the total estimated Spacecraft mass in terms of an initial estimated value, margin, and the total budgeted value for each subsystem. This table also includes the propellant mass estimate based on the propulsion analysis results in the table in Figure 7.4. The propellant mass to transfer to operational orbit uses the orbit parameters from the orbit analysis discussed previously and some assumed Launch Vehicle parameters [14]. It is clear from this example that there are many opportunities for on-going design trade studies that can result in updates to the required and estimated mass properties, other design properties, and orbit parameters. The integration of the analysis models with the system model is further summarized in the Perform Design Analysis activity in Chapter 9.

Subsystem	Initial Mass	Margin	Budget Mass
Payload	18.2 kg	7.8 kg	26.0 kg
Structure	16.1 kg	6.9 kg	23.0 kg
Thermal	1.4 kg	0.6 kg	2.0 kg
Power	13.3 kg	5.7 kg	19.0 kg
Communications	1.4 kg	0.6 kg	2.0 kg
Avionics	2.8 kg	1.2 kg	4.0 kg
GN&C	3.5 kg	1.5 kg	5.0 kg
Propulsion	2.1 kg	0.9 kg	3.0 kg
Other	2.1 kg	0.9 kg	3.0 kg
Propellant	35.0 kg	15.0 kg	50.0 kg
Total Dry Mass			87.0 kg
Total Wet Mass			137.0 kg
Launch Target Mass			150.0 kg
System Level Margin			13.0 kg

FIGURE 7.3

An initial system mass budget with launch mass target of 150kg based on the Pegasus Launch Vehicle [14].

Constants and Mission Parameters	
Nominal Dry Mass	100.0 kg
Insertion Orbit	200.0 km
Operational Orbit	700.0 km
Inclination Error	0.050 deg
Earth Radius	6378.0 km
Earth μ	398600.4418
I_{sp}	220.0 sec
Orbital Insertion	
Hohman Burn 1	270.2571 m/s
Hohman Burn 2	138.6632 m/s
Inclination Correction	6.5488 m/s
Propulsion Requirement	21.2285 kg
Yaw Slew	
Prop per maneuver	0.0490 kg
Maneuvers per year	10.0
Mission Duration	8.0 years
Propulsion Requirement	3.92 kg
Orbit Maintenance	
Drag	21.0 m/s
De-Orbit	183.00 m/s
Propulsion Requirement	9.9135 kg
Propulsion Budget Analysis	
Required Propulsion	35.0620 kg
Margin	11.5705 kg
Budgeted Propulsion	46.6325 kg

FIGURE 7.4

The Delta-V and Required Fuel Mass Estimate

7.5 Capture Spacecraft Requirements

The Spacecraft **black box specification** with each of its features is shown in Figure 7.5. The features of this block reflect what is required of the Spacecraft, and include the ports that specify the Spacecraft interfaces, the operations that specify the Spacecraft functions, the value properties that specify the Spacecraft performance, physical, and quality characteristics, and the state machine that specifies the Spacecraft states.

The goal of the integrated spacecraft design is to satisfy the requirements that are associated with each black box feature, and optimize the mission measures of effectiveness within the program constraints. Each system function is satisfied by functional threads resulting from interactions between the spacecraft components. The value properties, such as mass, power, and size are budgeted across the spacecraft subsystems and components.

Only critical performance, physical, and quality characteristics are shown in the black box. The keyword «*tpm*» identifies these system properties as **technical performance measures** (TPM) to indicate that they can significantly impact mission and system performance. In addition, the black box specification includes a reference to its Orbit, which specifies the Spacecraft orbit parameters. (Note that MoE applies to mission-level performance parameters and TPM applies to system level performance parameters.)

The black box specification can be generalized to specify a range of missions by specifying the more general functions, interfaces, properties, and states that are required to support s a set of missions. The more general black box specification can then be specialized for a particular mission by inheriting the common features and adding the unique features for the particular mission. The specialization also includes specifying the specific property values such as those for mass, power, and delta-v.

The spacecraft system requirements are captured in Figure 7.6 in a similar way that the mission requirements are graphically depicted. The system requirements are derived from the mission requirements based on analyzing the mission and determining what is required of the spacecraft to support the mission. These requirements are aligned with the features of the Spacecraft black box specification, often through a refine relationship. For example, the mass property of the black box specification refines the mass requirement in Figure 7.6, which means it is intended to re-express the text requirement in a more precise manner. (Note: An example is included in Figure 8.26).

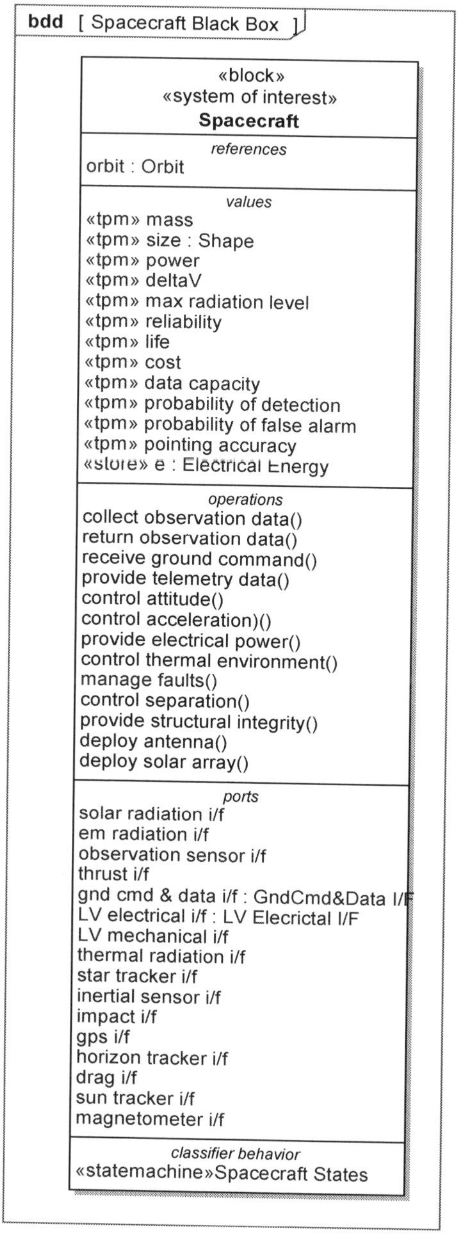

```
bdd [ Spacecraft Black Box ]

        «block»
    «system of interest»
        Spacecraft
─────────────────────────
           references
    orbit : Orbit
─────────────────────────
             values
    «tpm» mass
    «tpm» size : Shape
    «tpm» power
    «tpm» deltaV
    «tpm» max radiation level
    «tpm» reliability
    «tpm» life
    «tpm» cost
    «tpm» data capacity
    «tpm» probability of detection
    «tpm» probability of false alarm
    «tpm» pointing accuracy
    «store» e : Electrical Energy
─────────────────────────
           operations
    collect observation data()
    return observation data()
    receive ground command()
    provide telemetry data()
    control attitude()
    control acceleration)()
    provide electrical power()
    control thermal environment()
    manage faults()
    control separation()
    provide structural integrity()
    deploy antenna()
    deploy solar array()
─────────────────────────
             ports
    solar radiation i/f
    em radiation i/f
    observation sensor i/f
    thrust i/f
    gnd cmd & data i/f : GndCmd&Data I/F
    LV electrical i/f : LV Elecrictal I/F
    LV mechanical i/f
    thermal radiation i/f
    star tracker i/f
    inertial sensor i/f
    impact i/f
    gps i/f
    horizon tracker i/f
    drag i/f
    sun tracker i/f
    magnetometer i/f
─────────────────────────
         classifier behavior
    «statemachine»Spacecraft States
```

FIGURE 7.5

The Spacecraft black box specification specifies the key functional, interface, and technical performance measures the system design must satisfy to support the mission requirements. (Note: property values not shown)

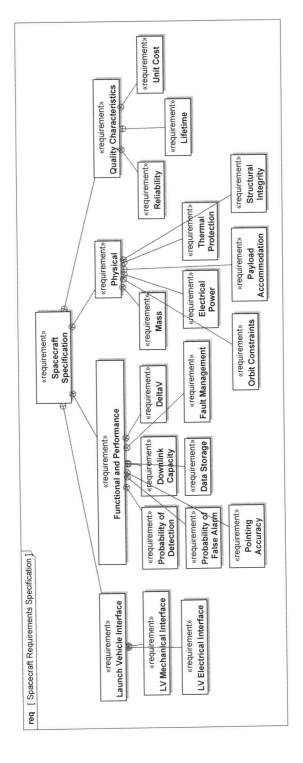

FIGURE 7.6

The Spacecraft Specification contains the text-based requirements that are refined by the black box specification features.

Figure 7.7 shows the derivation of the Spacecraft Mass and Delta-V requirement from the Launch Vehicle Compatibility, Number of Launch Vehicles, and Orbit requirements, and the rationale for that derivation based on the Mass and Delta-V Analysis, which is described in Chapter 9.2. Note that the rationale attaches to three derive requirement relationships.

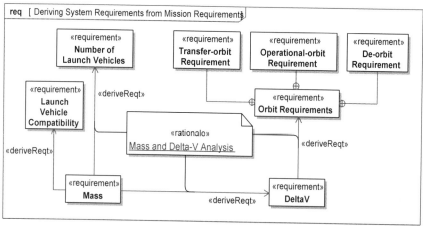

FIGURE 7.7

Example showing traceability of system requirements for Mass and Delta-V with supporting analysis providing rationale.

The Spacecraft Specification may be developed by the Contractor responsible for developing the Spacecraft, or provided by the Sponsor to the Contractor, or a combination of both. The Spacecraft Specification contains requirements that specify the Launch Vehicle Interface, Functional and Performance, Physical, and other Quality Characteristics. The Spacecraft requirements continue to be refined as part of the spacecraft design trade studies, to evolve a balanced system design that supports the mission objectives and associated measures of effectiveness.

SYNTHESIZE ALTERNATIVE ARCHITECTURES

CHAPTER 8

The purpose for this activity is to synthesize alternative architectures by partitioning the system that can satisfy the system requirements, and specify the components of a preferred architecture. This activity includes decomposing the black box spacecraft into its subsystems and components, and define their interconnection, interactions, and black box specifications. The component specifications are inputs to the component design and verification activities.

Architecture alternatives can be generated from scratch, but often are based on the experience of the architects and the development organization. There is increased emphasis on codifying the knowledge of an organization in the form or reference architectures, so that a new system design can leverage the previous knowledge in a systematic way. A reference architecture provides the basis for defining a family of solutions. To accomplish this, the reference architecture includes variation points that correspond to design choices, such as the number of reaction wheels, or different kinds of solar arrays or payload sensors. For a particular mission, selections from the design choices are made to satisfy the mission and system requirements. The architecture in this example can be viewed as a reference architecture that retains a set of variation points to accommodate a range of FireSat or similar remote sensing missions. This architecture is used as a basis for making architecture design choices to satisfy the specific FireSat II mission and system requirements.

This chapter addresses the following:

- Decompose the system into its components on a block definition diagram.
- Define the interconnection among the parts using the internal block diagram.
- Define the interaction among the parts using activity diagrams.
- Specify the components of each subsystem.

8.1 Define Spacecraft Decomposition and Interconnection

Spacecraft decomposition. There are many possible ways to partition the spacecraft into its components, although there are fairly standard system breakdowns that are used among the spacecraft design community. To develop the architecture, the Spacecraft black box specification is used as a starting point to decompose the Spacecraft into its subsystems and components. However, the Spacecraft black box specification shown in Figure 7.5 is not decomposed directly. Instead, the black box specification is specialized by another block that inherits the Spacecraft black box features, and this block is decomposed into its parts. This approach preserves the black box specification as different Spacecraft design alternatives are developed.

The Spacecraft black box specification can be specialized into a block that corresponds to a Spacecraft-Logical block and another block that corresponds to the Spacecraft-Physical block. The Spacecraft-Logical block is decomposed into logical components which do not impose technology-specific constraints. This is done when it is desired to open up the trade space to consider substantially new design alternatives. The logical components derived from the logical decomposition are allocated to alternative physical blocks that are used to synthesize alternative physical architectures. The logical decomposition is not included in this example primarily to reduce complexity of the approach. However, the physical components used in the reference architecture are sufficiently abstract to accommodate substantial variation in design options.

In Figure 8.1, the Spacecraft-Physical block is a specialization of Spacecraft black box specification from Figure 7.5 and inherits the black box specification features. The Spacecraft-Physical is decomposed into the Payload, Communications, Avionics, GN&C, Propulsion, Power, Thermal, Structure & Mechanisms, and Harness subsystems. The Software is also treated as a subsystem but is not shown in the figure. The *«subsystem»* key word designates the subsystems on the block definition diagram.

The Spacecraft subsystems are further decomposed into their parts as shown in their parts compartments. The parts that are considered to be logically part of a subsystem, are shown in the subsystem's references compartment. For example, the Power Subsystem contains the Array Deployment Mechanism and the Solar Array Gimbal in its reference compartment, but these are also shown as parts of the Structure & Mechanisms Subsystem, which is considered the owner of these parts.

The use of multiplicity enables the specification of variant Spacecraft designs in many of the subsystems. For example, the number of Solar Arrays varies from 1 to many, and the number of reaction wheels varies from 0 to 3. These design choices are subject to further trade study and analysis. As will be shown later, multiplicity and specialization provide substantial capability to model variation points that identify design choices for alternative physical architectures.

The software components are treated as a part of the subsystem that they support. For example, the Communication Subsystem owns the Communication Software. As described in Section 8.2.9, the software components are aggregated into a Software Subsystem. There are other ways to represent the software in the Spacecraft system hierarchy. For example, the software components could be part of the Software Subsystem, and referenced by the other subsystems that they are logically associated with. The approach should be defined and applied consistently throughout the system model.

The Spacecraft design may have multiple decompositions to support different views of the product structure, but a single decomposition should be designated as the primary decomposition. Each decomposition requires a specialization of the Spacecraft black box specification. An example is a decomposition that aggregates the components into higher level assemblies that generally do not correspond to subsystems. For example, each Remote Interface Unit (RIU) in the Command & Data Handling (C&DH) subsystem may be a part of a different assembly. A production view of the system typically includes a system decomposition into assemblies that correspond to the way the Spacecraft is physically assembled on the manufacturing floor. Another example decomposition is a dynamics view that decomposes the Spacecraft into lumped masses that include the center body and appendages such as arrays, antennas, and thrusters. The lumped masses include their mass, moment of inertia, and other mass properties, and are subject to dynamics analysis. The correspondence between the lumped masses in the dynamics decomposition and the components in the primary decomposition must be maintained, and the constraints on the property values must be consistent (i.e., the mass of the center body must equal the total mass of all of the components that correspond to the center body). A third example is a decomposition that is more aligned with a work breakdown structure (WBS). In this example, the components of the system may be aggregated based on the organization that is responsible for them.

Spacecraft subsystem interconnection. The Spacecraft Subsystem Interconnection is shown in Figure 8.2. The Spacecraft-Physical is the enclosing block for this internal block diagram, and the parts correspond to the subsystems in the block definition diagram in Figure 8.1. The connectors connect the ports on each subsystem. In this view, the connectors between the Power, Thermal, and Structure & Mechanisms subsystems and the other subsystems are not shown to simplify the diagram. However, the model does include the connectors that represent different kinds of interconnection, including command and telemetry, power, thermal, and structural connections.

Some item flows are shown such as the input Emissions to the Payload Subsystem, the RF signals to and from the Communications Subsystem that correspond to the uplink ground commands and the downlink observation and telemetry data, and the digital messages that flow between the Avionics Subsystem and Communications Subsystem. The spacecraft includes a diverse range of interfaces and item flows that include both physical and information flows. The representation of the interfaces in the model can vary from highly abstract, such as named ports, to highly detailed, such as a model of a layered communication protocol stack [13].

ARCHITECTING SPACECRAFT WITH SYSML

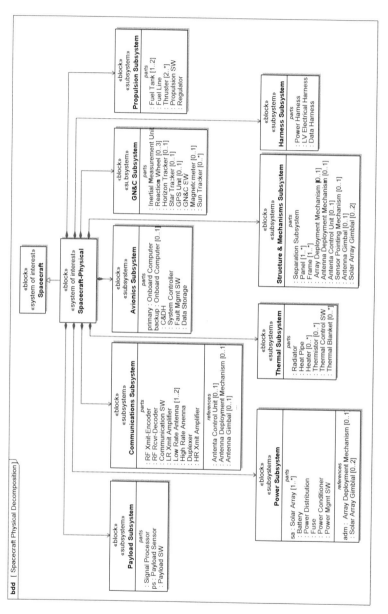

FIGURE 8.1

The Spacecraft-Physical is a specialization of the Spacecraft black box specification, and is decomposed into its subsystems and their parts.

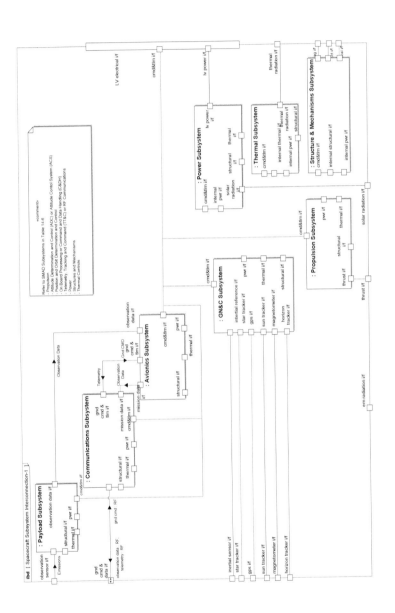

FIGURE 8.2

The internal block diagram for the Spacecraft-Physical shows the interconnection between the subsystems. The connectors to the Power, Thermal, and Structure & Mechanisms subsystems are not shown to simplify the diagram.

8.2 Identify and Evaluate Subsystem Design Alternatives

There are many spacecraft design alternatives to be considered throughout the development process. The following sections describes each subsystem, identifies some of the design trades, and selects a preferred solution based on the design selections make in the New SMAD FireSat II example. The Power Subsystem section includes an example of a Solar Array trade study using the system model to specify the alternative design concepts, and analysis models to evaluate the alternative designs. This can be used as a representative example for how other subsystem trade studies can be performed.

8.2.1 Payload Subsystem

The Payload Subsystem interconnection diagram is shown in Figure 8.3. This subsystem provides the primary mission functionality to support the mission. For the FireSat II mission, the sensor must be capable of detecting temperature variance on the earth. In addition to this sensor, the Payload Subsystem includes a Signal Processor which post-processes the observation data that is later downlinked to the ground. The Payload SW controls the Payload Sensor and provides payload telemetry.

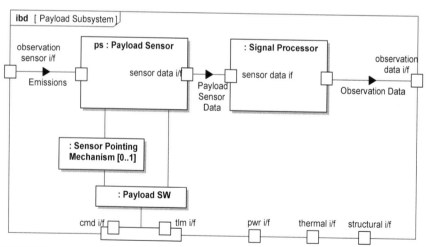

FIGURE 8.3

The Payload Subsystem senses and processes the thermal emissions from the subject region. This subsystem includes a Payload Sensor, Signal Processor and optional Sensor Pointing Mechanism, and Payload SW to control the sensor.

If required, this subsystem can include a Sensor Pointing Mechanism. The [0..1] multiplicity indicates that a valid solution can contain either zero or one of these mechanisms. The dashed outline indicates that this mechanism is referenced by the Payload Subsystem, but is owned by another subsystem (i.e., Structure & Mechanisms Subsystem).

There are several trade studies to be performed to select the preferred Payload design. The type of sensor must satisfy the payload performance requirements and mass and power budgets. It also must be determined whether a pointing mechanism is required to point the sensor, or whether this is accomplished by pointing the Spacecraft at the earth.

The analysis indicates that the preferred selection for the Payload Sensor is a Mid-Range IR Scanner which is mounted directly to the body of the Spacecraft. The Spacecraft is nadir pointing, and therefore the pointing of the sensor is controlled by pointing the Spacecraft. This sensor generates data at a rate of 8 megabytes per second (MBPS). This data rate drives an additional trade study between downlink capability and on-board data processing capability. The downlink rate must be sufficient to downlink all data to the Ground Station, or additional data processing must be performed on-board the Spacecraft to reduce the amount of data to be sent – perhaps only sending data frames when a fire is detected.

8.2.2 Communications Subsystem

The Communications Subsystem interconnection diagram is shown in Figure 8.4. This subsystem provides capability to communicate with the FireSat II Ground Station and supports uplink of vehicle commands, downlink of telemetry data and downlink of Payload Sensor observation data. The Spacecraft must communicate two different kinds of data with the ground each with differing data rate requirements. As such, this requires two data channels within the Communications Subsystem – one low rate for commands and telemetry and one high rate for observation data.

This subsystem includes one High Rate Antenna and either one or two Low Rate Antennas as indicated by the multiplicity in the figure. The low rate channel supports both uplink and downlink and therefore has both a receiver decoder and a transmit encoder. The receiver decoder processes incoming command data and broadcasts it to the cmd i/f of the subsystem. Status data flowing across the tlm i/f is processed by the communication software prior to being encoded and amplified for transmission.

The Low Rate Antenna design is fixed and not deployed. The high rate channel has an optional Antenna Deployment Mechanism (such as a hinge), an Antenna Control Unit and Antenna Gimbal to allow for antenna steering if necessary. This can increase communication capacity with the Ground Station, but adds additional cost and complexity to the design.

ARCHITECTING SPACECRAFT WITH SYSML

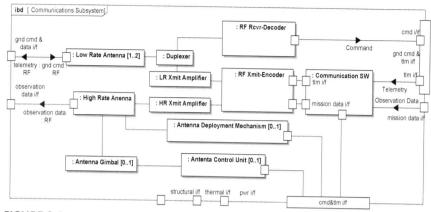

FIGURE 8.4

The Communication Subsystem includes a low rate channel to support the uplink of commands and downlink of telemetry, and a high rate channel to support the downlink of the observation data.

The selection of the transmit frequency is an important design decision. The transmit frequency must accommodate the Payload Sensor data rate that generates data at 8 MBPS. This design alternative does not include on-board processing to reduce the downlink data, nor does the mission configuration leverage a relay satellite for ground communication. As a result, the Communication Subsystem must be capable of downlinking all data collected in an orbit in the 10 minutes (i.e. approximately 1/10th of the orbit period) that the Spacecraft is in contact with the Ground Station. This requires the high rate channel to support downlink at least 80 MBPS, resulting in the selection of an X-band transmitter for the high rate channel.

8.2.3 Avionics Subsystem

The Avionics Subsystem interconnection diagram is shown in Figure 8.5. This subsystem provides the on-board processing and command and data handling (C&DH) for the Spacecraft. The C&DH can be thought of as a data router where command and telemetry data are routed to the appropriate component or subsystem for processing. The ports on the left side of the Avionics Subsystem provide the interface to the Communication Subsystem to receive commands from the ground and send telemetry to the ground. The ports on the right side of Avionics Subsystem provide the interface to send commands and receive telemetry to and from the other subsystems.

The System Controller is responsible for controlling the state of the Spacecraft by monitoring the current subsystem state and state changes, and sending commands to trigger state changes of one or more other subsystems. The Fault Management Software provides additional monitoring and control

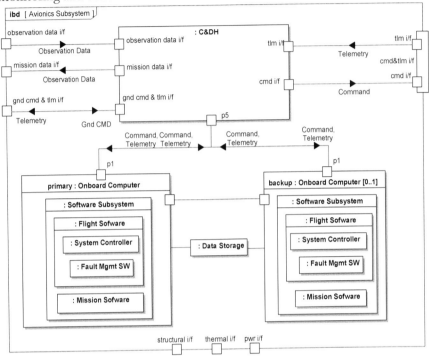

FIGURE 8.5

The Avionics Subsystem provides the on-board data processing and routes the data to other Subsystems.

of system health, and triggers responses to mitigate failures as part of the overall fault detection, isolation, and recovery (FDIR) approach.

The on-board processing for FireSat II includes two on-board computers that execute the system software and provide the on-board Data Storage. One computer acts as primary, and the other as backup. Duplicate on-board computers increase system reliability, but at increased cost. If it is determined that the reliability of a single on-board computer is sufficient, the second computer can be removed.

The Avionics Subsystem includes a writeable data storage device to store Payload Sensor data for downlink. As noted previously, the Payload Sensor generates data at 8 MBPS, and the mission orbit requires 99 minutes for a full revolution. As such, the data storage device must be sized to store at least 50GB of data and have a write rate of at least 8 MBPS. This storage is implemented in flash memory to meet the mass, power and reliability requirements as well as the data requirements.

8.2.4 Guidance, Navigation, and Control (GN&C) Subsystem

The Guidance, Navigation and Control (GN&C) Subsystem interconnection diagram is shown in Figure 8.6. This subsystem measures the state of the Spacecraft and controls its attitude. The GN&C Subsystem is comprised of a combination of sensors to determine where the Spacecraft is and where it is pointing, which may include an earth Horizon Tracker, a Sun Tracker, an Inertial Measurement Unit, and a GPS unit. Additionally, the GN&C Subsystem includes Reaction Wheels to allow for attitude adjustment of the Spacecraft with minimal use of fuel. These wheels are required to maintain the pointing angle of the sensor, and to perform the "yaw flip" maneuver to maximize exposure of the solar array to the sun.

The GN&C Subsystem is a closed loop control system which makes many measurements to understand the current state of the Spacecraft. It has a significant amount of software to determine how to respond to the current Spacecraft state. For example, when the Spacecraft requires a "yaw flip" maneuver, the GN&C Subsystem must either spin up or slow down the Reaction Wheels to torque the vehicle, and collect data from Sun Sensors to determine when the maneuver is complete. It must also command the Propulsion Subsystem to dump excess momentum.

There are various sensor types that may be used to measure the vehicle pointing angle, inertia, momentum and location based on the Spacecraft mission need. For instance, in a low earth orbit, GPS may be used to determine the current vehicle location to determine if an orbit adjust maneuver is necessary. This data could supplement tracking data from the ground station to provide a higher accuracy assessment of vehicle position.

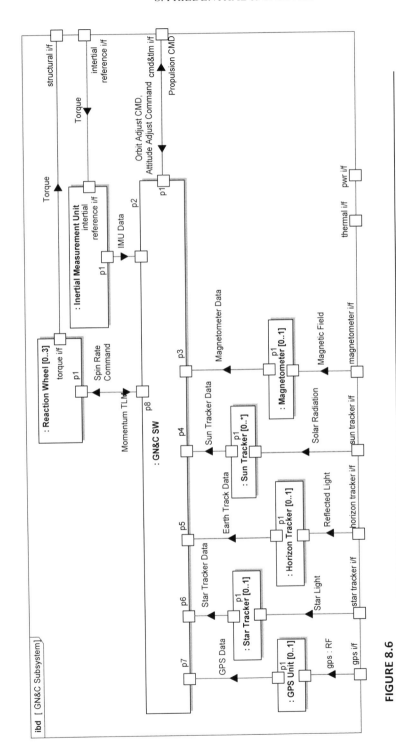

FIGURE 8.6

The GN&C Subsystem includes multiple sensors to measure the state of the Spacecraft and control its attitude.

8.2.5 Propulsion Subsystem

The Propulsion Subsystem interconnection diagram is shown in Figure 8.7. This subsystem is responsible for providing thrust to affect the delta-v of the Spacecraft. This subsystem is composed of one or more Propellant Tanks, Regulators, Pressurant Tanks, Propellant Lines, and multiple Thrusters. The number of Thrusters is subject to trade study. The design selection for this example includes two Thrusters per controlled axis (i.e., pitch and yaw). The Propulsion SW determines the duration of fuel flow to the Thruster in response to a propulsion command.

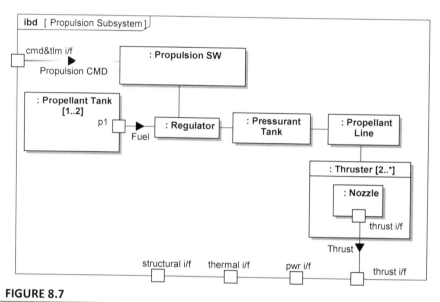

FIGURE 8.7

The Propulsion Subsystem generates thrust to affect the delta-V.

A primary Propulsion Subsystem trade study is the selection of a mono-propellant or a bi-propellant system. A mono-propellant system includes a single tank to hold the propellant such as hydrazine, and can generate thrust without the use of an oxidizer. A bi-propellant system includes fuel and an oxidizer which are combined in a reaction chamber within the Thruster to generate thrust. A higher specific impulse can be achieved from a bi-propellant system, but bi-propellant designs are more complex. They require multiple Propellant Tanks, separate lines for fuel and oxidizer, and more control software. Monopropellant thrusters can produce a specific impulse of 220 seconds. The black box propulsion analysis indicated this specific impulse is sufficient to support the required maneuvers, so the preferred option is the simpler mono-propellant design.

8.2.6 Thermal Subsystem

The Thermal Subsystem interconnection diagram is shown in Figure 8.8. This subsystem is responsible for maintaining the thermal environment within the Spacecraft. The Thermal Subsystem is composed of temperature sensors (i.e., Thermistors), Heaters, Thermal Blankets, a Radiator, Ammonia filled Heat Pipes, and Thermal Control Software.

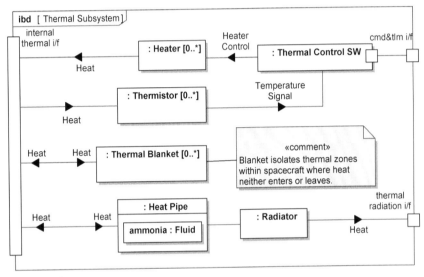

FIGURE 8.8

The Thermal Subsystem controls the temperature across the Spacecraft, and its components are physically distributed.

The number of Heaters and Thermistors are heavily dependent on the design of the Spacecraft. Each component in the system has a nominal temperature operating range. The Thermistors are placed on or near those components so that their temperature can be monitored. The Heaters are mounted to ensure component temperatures stay within their nominal operating range. Thermal blankets insulate components and fuel to maintain a stable temperature environment.

Since the Spacecraft is nadir pointing, there is a significant temperature gradient across the Spacecraft. The side of the Spacecraft pointing at the Sun will be hotter than the opposite side. The IR Payload Sensor and Radiator must be kept cool, since the Spacecraft is not spun to even out the temperature. Therefore, the Heat Pipes and a Radiator are important elements of the thermal control approach. Heat pipes absorb heat from the hot side of the Spacecraft and circulate ammonia to transfer the heat to the Radiator where it can be expelled from the spacecraft. Balancing the temperature across the Spacecraft requires significant analysis and iteration with the physical layout.

8.2.7 Power Subsystem

The Power Subsystem interconnection diagram is shown in Figure 8.9. This subsystem provides the electrical power for the Spacecraft. It includes the Solar Array to generate the electrical power, the Battery to store the energy, and the Power Conditioner, Power Distribution, Power Switch, and Power Management Software. Prior to separation, the Launch Vehicle provides the electrical power to the Spacecraft. Following separation, the Spacecraft relies on its batteries for its electrical power, and the Solar Array to maintain the battery charge.

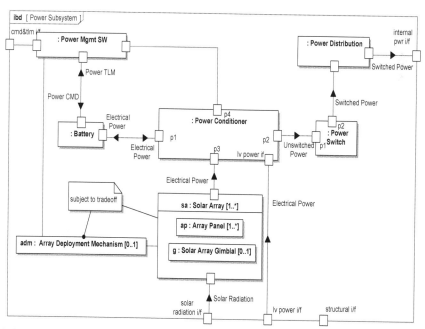

FIGURE 8.9

The Power Subsystem provides the electrical power for the Spacecraft, and includes a Solar Array to generate the power and Battery to store the energy.

The power and/or energy budget analysis is dependent on the type of mission or mission phase, and may be based on a combination of steady state power and peak power demands. The required solar power that the solar arrays must collect from the sun must be sufficient to charge the batteries and satisfy the power consumption demand of the Spacecraft. The orbit analysis is used to establish the power consumption scenarios that define when the Payload Sensors are operating, when the Spacecraft is downlinking data to the Ground Station, and when Heaters are on. The initial power consumption estimate is based on these scenarios and comparisons with similar designs. The preliminary power budget is shown in the table in Figure 8.10 along with some of the assumptions.

Subsystem	Estimated Power (W)	Margin (W)	Budget Power (W)
Payload	45	20	65
Structure	1	0	1
Thermal	10	4	14
Power	9	4	13
Communications	12	5	17
Avionics	12	5	17
GN&C	10	4	14
Propulsion	0	0	0
Estimated Power Consumption			141 W
Battery Efficiency			98 %
Xe Factor			0.97
Te Factor			0.36
Xd Factor			0.99
Td Factor			0.64
Margin			2 W
Total Solar Array Power Required			235 W

FIGURE 8.10

An initial power budget analysis and power requirement assessment.

A key trade study is the selection of the Solar Array configuration. The Solar Arrays are most efficient when they are pointed directly at the sun. The different design alternatives include body mounted arrays and gimbaled arrays. The latter provides increased efficiency since it can be pointed more directly at the sun. The gimbaled arrays must be deployed after separation, whereas the body mounted arrays are available immediately upon separation. The system model provides a means to capture the alternative configurations, the assumptions, and other inputs needed to perform the analysis. The parametric model for this analysis is shown in Figure 9.5. The following provides a brief summary of the trade study analysis.

The alternative Power Subsystem designs include a body-mounted solar array, a single-axis gimbal solar array ,and a two-axis gimbal solar array. The body-mounted and two-axis gimbal alternative decompositions are shown in the block definition diagram in Figure 8.11.

ARCHITECTING SPACECRAFT WITH SYSML

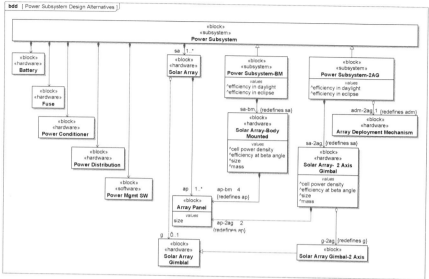

FIGURE 8.11

Alternative Power Subsystem designs include a body mounted Solar Array and 2 Axis Gimbal Solar Array.

The alternative designs for the Solar Array-Body Mounted and Solar Array-2 Axis Gimbal are shown as specializations of the generic Power Subsystem. Each alternative inherits the common parts from the generic Power Subsystem, and adds their unique parts via their decomposition. The body-mounted alternative contains 4 Array Panels, and the two-axis gimbal alternative contains 2 Array Panels, a two-axis gimbal, and an Array Deployment Mechanism.

The trade study results from the analysis of the three design alternatives are shown in the table in Figure 8.12 including consideration for power, mass, reliability, resilience, and cost. The analysis results indicate that the body-mounted configuration is the preferred solution.

	S/A Efficiency	Required Area (m2)	Actual Area (m2)	Mass (kg)	Total Cost ($)	Reliability Factor	Resiliency Factor	Objective
Power - Body Mount	0.6	1.99	2.02	5.66	30,240	1.00	0.8	**0.8283**
Power - One Axis	0.75	1.58	1.58	4.43	38,327	0.90	0.9	0.6471
Power - Two Axis	0.95	1.26	1.26	3.51	47,556	0.80	1	0.3898

SA Capacity (W/m2)	197
Required Power (W)	235

FIGURE 8.12

Solar Array trade study results.

This trade investigates the required solar array area for each design alternative, and its efficiency in collecting solar energy. An array with a two-axis gimbal can be controlled so that the array directly points at the Sun most of the time, providing an efficient design. A single-axis gimbal cannot always point directly at the Sun, but can effectively cut the angle between the spacecraft and the Sun. The orbit analysis models have been used to determine the worst case angle between the Spacecraft orbital plane and the sun vector. (i.e., beta angle) and has been used to estimate the improvement in angle a single axis solar array can achieve. Further, both the single and dual axis solar arrays will themselves require power to manage the array, further lowering the efficiency of each. This analysis has resulted in estimated efficiency for the single axis design of 75% and 95% for the two axis design.

Finally, the body mounted array does not have the flexibility to point the arrays at the Sun like the other two designs. Given the shape of the spacecraft, there are different efficiencies on each face of the spacecraft throughout the year. A conservative estimate based on analysis is that the body mounted array can meet a 60% efficiency as compared to the two-axis gimbal. The efficiencies are used to calculate the required Solar Array area for each design alternative. In the case of the body mounted design, the area is assumed to be a multiple of the size of a side of the Spacecraft.

Given the Solar Array size, the mass, cost, reliability, and resiliency of each design alternative is evaluated. The body-mounted design requires the largest total area, but is the least expensive design alternative since the cost of the Solar Array material is less than the additional cost of the gimbal mechanisms and control software. Additionally, the reliability factor for this body-mounted array design is higher since there are fewer moving parts. Finally, the larger subsystem mass allocation for the body-mounted design is still within the initial mass budget. As a result of the analysis and assumptions, the body-mounted design alternative is the preferred design solution.

The model provides considerable ability to specify variation and alternative design options using a combination of specialized blocks, such as the body-mounted Power Subsystem and the two-axis gimbal Power Subsystem, and the use of multiplicity, such as the number of array panels, and the optional multiplicity on the number of gimbals and the Array Deployment Mechanism.

8.2.8 Structure & Mechanisms Subsystem

The Structure & Mechanisms Subsystem interconnection diagram is shown in Figure 8.13. This subsystem provides the physical support infrastructure for the Spacecraft, structural panels for mounting the Solar Array and internal components, and the mechanisms to control movement of other subsystem components, such as the Antenna Gimbal, Antenna Deployment Mechanism, and the Separation Mechanism to separate from the Launch Vehicle.

The mechanisms are referenced by other subsystems. Mechanisms are mechanized joints between the rigid body of the spacecraft and a movable component such as the deployable antenna. The mechanism requirements are typically driven by the design of other subsystems such as the required antenna pointing accuracy for the Antenna Gimbal to maintain a link budget.

The structure is comprised of a Spacecraft Frame, a number of Panels and a mechanical interface to the Launch Vehicle and its faring. The number of Frames is influenced by the internal volume required for components and the size of the Launch Vehicle faring. The Separation Mechanism and Launch Vehicle interface depend on the selection of the Launch Vehicle. The structure and other aspects of the mechanical design of the Spacecraft are captured in a 3D CAD modeling tool.

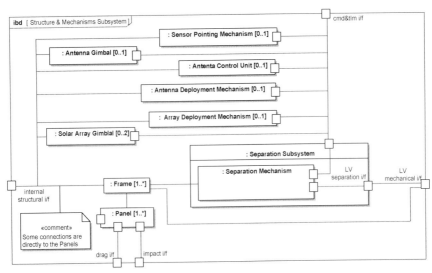

FIGURE 8.13

The Structure & Mechanisms subsystem includes the physical support for the Spacecraft and the moving parts.

8.2.9 Software Subsystem

The software components interact with the hardware components to provide much of the Spacecraft functionality. The software components are part of the other subsystems as described in Section 8.1 and shown in the subsystem internal block diagrams in this chapter. They may execute on a single mission computer, or execute on many processors distributed across the spacecraft. Each component can be allocated to the processor that executes it, but this design decision has been deferred in this example.

It is sometimes useful to treat the collection of software components as a subsystem. The Software Subsystem is shown in Figure 8.14. The Software Subsystem is decomposed into Flight Software and Mission Software. The Flight Software references the System Controller, Fault Manager SW, and GN&C SW, and the Mission Software references the software components for the other subsystems. The interaction between software components, such as the processing of commands and telemetry or fault management scenarios, are often represented using sequence diagrams.

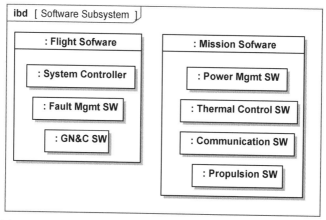

FIGURE 8.14

The Software Subsystem references the software components that are part of other subsystems, such as the Power Mgmt SW which is part of the Power Subsystem.

8.2.10 Harness Subsystem

The Harness Subsystem is the physical medium that connects electrical components to support the exchange of electrical data and the distribution of electrical power. The Harness Subsystem design must support the signal frequencies and data rates, and power levels, and consider the effects of attenuation, shielding, grounding, and many other characteristics that impact electrical interconnection.

In the subsystem interconnection diagram in Figure 8.2, the Harness Subsystem is abstracted away to simplify this interconnection view. The connectors correspond to logical connections between ports on the subsystems. The data and power ports can contain nested ports to represent specific electrical pins, enabling the specification of the pin-to-pin interconnection. If desired, the individual cables and harnesses can be represented as blocks and treated like any other component. These blocks can have properties such as mass, attenuation, and shielding characteristics, as well as internal structure such as harness segments.

8.3 Allocate Functional & Performance Requirements

Spacecraft system functional threads. Each system function in the black box specification in Figure 7.5 can be further elaborated by a system functional thread that specifies what each component must do to perform the system function. Each functional thread is expressed as an activity diagram, sequence diagram, or combination of the two. Figure 8.15, Figure 8.16, and Figure 8.17 show activity diagrams for selected Spacecraft functions to Provide Electrical Power, Control Attitude, and Collect and Downlink Observation Data.

The Provide Electrical Power activity shows the high level actions that Avionics Subsystem, the Power Subsystem, and the Power Harness perform to deliver electrical power to the subsystems. The inputs include Solar Radiation to the Solar Array, and the Electrical Power provided by the Launch Vehicle prior to Spacecraft separation. (Note: Generate System Commands includes some outputs not used in this activity, but used in other activities).

The Control Attitude activity shows the high level actions that the Avionics, GN&C, and the Propulsion Subsystems perform to control the attitude of the Spacecraft. This is accomplished by sensors that measure the sun angle, earth horizon angle, and the Spacecraft angular rate to control the Reaction Wheel spin rate and the thrust. The GPS and Star Tracker are left off for brevity.

The Collect and Download Observation Data activity shows the high level actions that the Payload, Avionics, and Communication Subsystems perform to sense the thermal emissions from the Earth, store the data, and then downlink the data to the ground. This activity shows the input and output as streaming, which means the input and output are consumed and produced by the activity as it executes, where-as a non-streaming input and output are only consumed and produced at the start and end of execution. The inputs and outputs to the activities in Figure 8.15 and 8.16 are also streaming.

ARCHITECTING SPACECRAFT WITH SYSML

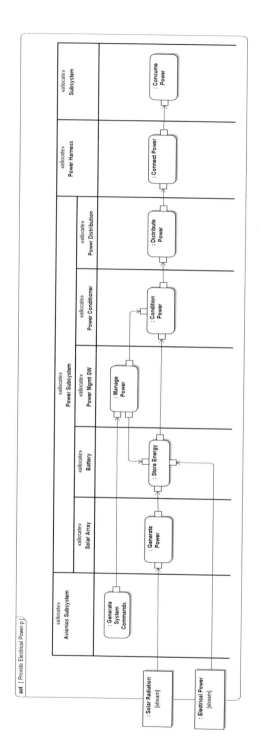

FIGURE 8.15

Provide Electrical Power activity diagram.

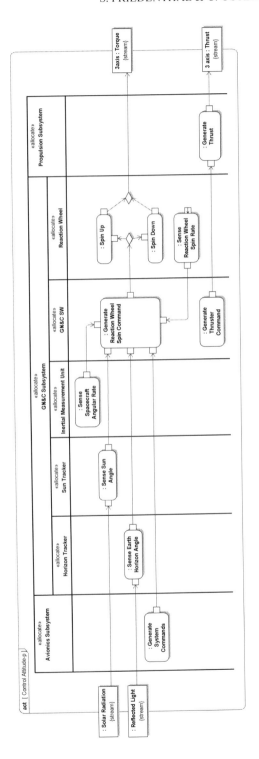

FIGURE 8.16 Control Attitude activity diagram.

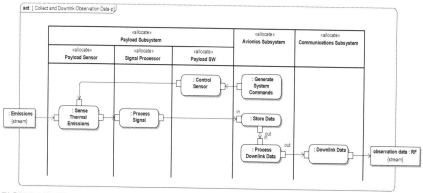

FIGURE 8.17

Collect and Downlink Observation Data activity diagram.

The activities can be further refined, and then elaborated by showing the actions performed at the next level of decomposition. For example, the Downlink Data action performed by the Communication Subsystem in the Collect and Download Observation Data activity can be further decomposed into another activity with actions that the components of the Communication Subsystem perform, including packaging, encoding, and amplifying the data, and then transmitting the data via the High Rate Antenna. These activities can be executed by the system modeling tool simulation engine to facilitate further analysis of the input-output and control flow.

Spacecraft event timeline. The mission timeline shown in Figure 6.12 and Figure 6.13 identifies key mission events throughout the spacecraft mission, beginning with launch, transfer to orbit, on-orbit operations, and de-orbit. More detailed timelines of the Spacecraft internal events are critical to the Spacecraft design and analysis.

The sequence diagram in Figure 8.18 highlights a simplified example of a portion of the Spacecraft timeline, indicating the events and time durations that the Payload Sensor is collecting data, the Transmitter is amplifying signals to downlink, and the Solar Array is exposed to the sun. The events and time durations are used to specify power analysis scenarios, and can be expanded to include the other electrical components that consume power.

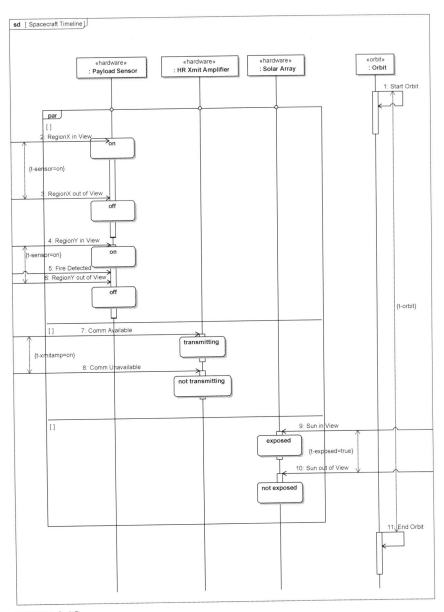

FIGURE 8.18

The Spacecraft timeline captures internal events such as the Payload Sensor turning on or off. These timelines represent scenarios which are integrated with simulation and analysis tools to support various analysis, such as power analysis.

Component failure modes. Failure modes are defined as a violation of nominal or acceptable performance of an activity or function. A failure mode can cause another failure mode. For example, the mission failure modes in Figure 6.14 include a failure mode that occurs when the Control Trajectory activity results in a trajectory that is outside it's acceptable bounds. Figure 8.19 shows two attitude control failure modes that can cause a Trajectory Failure, including both a Steady State Attitude Control Failure and an Attitude Rate Control Failure. The Attitude Rate Control Failure can fail when the attitude rate is too high or too low.

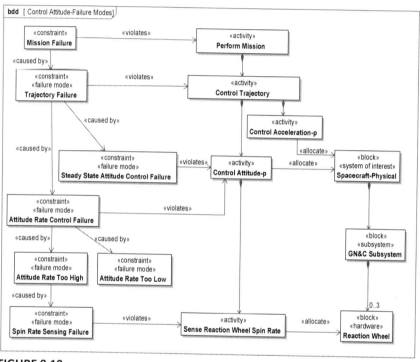

FIGURE 8.19

Failure modes are related to activities (i.e., functions), components, and other failure modes.

As further indicated in this figure, an excessive attitude rate can be caused by a Spin Rate Sensing Failure. This failure mode corresponds to the Reaction Wheel's inability to perform its Sense Reaction Wheel Spin Rate function. In this way, the functional and associated component failure modes can be systematically identified and input to the fault tree analysis and the failure modes, effects, and criticality analysis (FMECA).

The strategy to mitigate these failures includes increasing design reliability through parts selection, redundancy, simplicity, and other means. This is coupled with an effective fault detection, isolation, and recovery (FDIR) approach that includes fault containment, error monitoring, and mitigating responses.

Component states. The states for each component can be specified by their state machine. The electrical components include an off and on state. Some components have many states, such as the System Controller that is used to control the behavior of the Spacecraft. The System Controller state machine must result in system states that are consistent with the Spacecraft state machine specified in Figure 7.2.

The Spacecraft operates in the Normal state when it is within its nominal performance bounds. The Fault Management Software is responsible for controlling the response to off-nominal behavior. The state machine for the Fault Management Software is shown in Figure 8.20. The Fault Management Software responds to signals from error monitors. If an error is detected, the Fault Management Software requests a local response. If the error persists, the Fault Management Software requests a system response such as turning a device off and transitioning to a backup device. If a mission critical threshold is exceeded, the Spacecraft transitions to the Safe state, where additional autonomous actions, or actions commanded from the ground, may be performed to try to return to the Normal state.

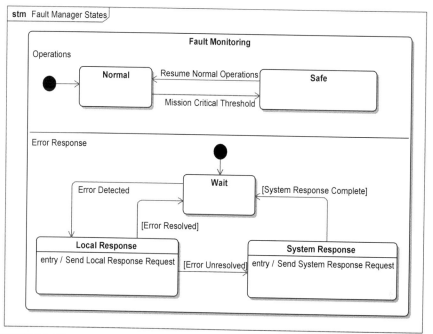

FIGURE 8.20

Fault Management States include Fault Monitoring states and Error Response states.

Component performance requirements. The critical component properties and their values are captured as value properties on each component block. The values are derived from analysis, such as the power and mass analysis described previously. The governing equations can be captured in parametrics, and integrated with analytical models as described in Chapter 9. The property values can also vary with its component's state. For example, the power consumption for the Payload Sensor is shown in its on and off states in Figure 8.21, and the power received from the Sun is shown in the Solar Array states in Figure 8.22.

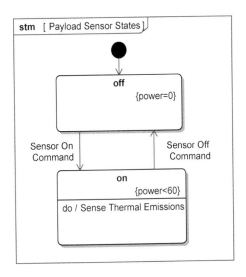

FIGURE 8.21

The required power consumption (i.e. power) of the Payload Sensor in the on state is captured in its state machine.

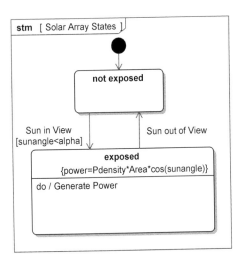

FIGURE 8.22

Solar Array states with constraints on power.

8.4 Capture Component Requirements

The component requirements are captured as black box specifications and text based requirements similar to how the system requirements are captured.

Component black box specifications. The subsystems and components can be specified as a black box in the same way as the system black box specification is defined in Figure 7.5. An example is the black box specification for the Payload Sensor shown in Figure 8.23. As with the system black box, the Payload Sensor black box includes its interfaces, functions (i.e. operations), critical performance, physical, and quality characteristics, and its state machine. The constraints on some of its critical properties reflect its requirements.

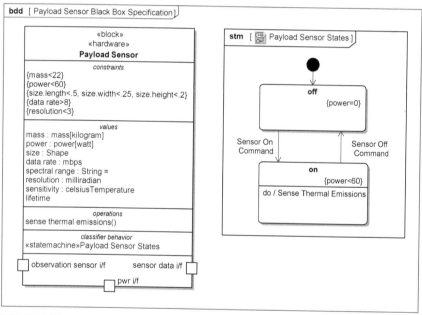

FIGURE 8.23

The Payload Sensor black box specification can be realized by alternative design solutions such as a Mid-Range IR Scanner.

Component requirements specifications. The component requirements can be specified in terms of traditional text requirements similar to the system requirements specification shown in Figure 7.6. The Payload Sensor Specification is shown in Figure 8.24 and Figure 8.25 in both graphical and tabular form, respectively.

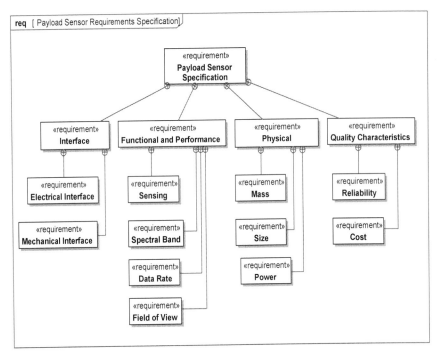

FIGURE 8.24

The Payload Sensor Specification contains requirements that align with the black box specification.

#	Id	Name	Text
1	80	Payload Sensor Specification	
2	80.1	Quality Characteristics	
3	80.1.1	Reliability	
4	80.1.2	Cost	
5	80.2	Physical	
6	80.2.1	Mass	The sensor mass shall be less than 22 kg.
7	80.2.2	Size	The sensor size (length, width, height) shall fit in a rectangular volume of less than 0.5, 0.25, 0.20 metres
8	80.2.3	Power	The sensor power consumption shall be less than 60 watts ir the on state.
9	80.3	Functional and Performance	
10	80.3.1	Sensing	The sensor shall sense thermal emissions with a sensitivity less than TBD, and a resolution of less than 3 miliiradians.
11	80.3.2	Field of View	The sensor shall be able to scan a field of view of at least TBD milliradians.
12	80.3.3	Data Rate	The sensor shall be capable of providing a data rate of a minimum of 8 Megabits per second.
13	80.3.4	Spectral Band	The spectral band shall be mid range infrared
14	80.4	Interface	
15	80.4.1	Electrical Interface	
16	80.4.2	Mechanical Interface	

FIGURE 8.25

A tabular view of the preliminary Payload Sensor Specification that is shown graphically in the previous figure.

The selected sensor for this design that satisfies the black box specification is a Mid-Range IR Sensor. This sensor inherits the features of the black box specification, but includes the design values for its properties as indicated in Figure 8.26. Other information, such as its designation as a Commercial off-the-shelf (COTS) item versus a development item, can also be specified using customized properties.

The traceability relationships to the text requirements in the Payload Sensor Specification are maintained. The features of the Payload Sensor black box specification, including its value properties, operations (i.e., functions), and ports, refine the text requirements. The corresponding features of the IR Scanner satisfy the text requirements. An shown in Figure 8.26, the mass property of the black box specification refines the Mass requirement, and the mass property of the IR Scanner satisfies the Mass requirement. These relationships are captured in requirements tables (not shown) as a more compact representation. Examples of other traceability relationships are discussed in Chapter 10.

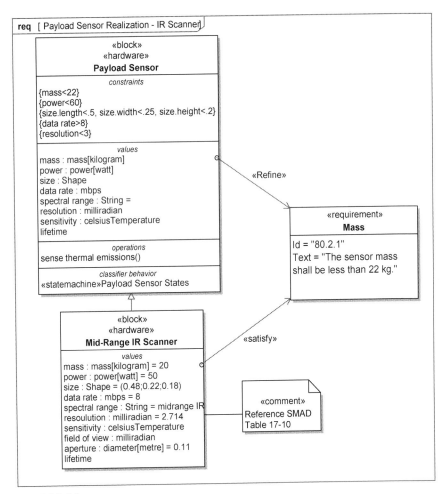

FIGURE 8.26

The mass property of the Payload Sensor black box specification refines the Mass requirement and the mass property of the Mid-Range IR Scanner satisfies the Mass requirement of the Payload Sensor Specification.

8.5 Capture Preferred System Design Configuration

The selection of a preferred system design is a highly iterative process involving the identification and analysis of design alternatives. The method for specifying the analysis to be performed is described in the next chapter. This iterative process is intended to yield a balanced architectural solution and the associated component requirements that satisfy the spacecraft mission and system requirements.

The preferred FireSat II system design includes the selection among many different subsystem design alternatives that were identified as variation points of the more general spacecraft reference architecture. Some of the design choices for the preferred spacecraft design choices that are described in this chapter include a body-mounted solar array, monopropellant propulsion, 2 thrusters per axis, a low rate and deployable X band high rate antenna, and the fixed Mid-Range IR Scanner. This preferred system design configuration is captured in the model and put under baseline control.

The geometric model is developed concurrent with the system model to provide a geometric view of the spacecraft. This model is typically captured in a 3 dimensional computer-aided design (CAD) model using commercial tools. A conceptual design of the top assembly view of the preferred spacecraft design alternative is shown in Figure 8.27 and Figure 8.28. Figure 8.27 shows the low and high rate antennas, Mid-Range IR Scanner, and the Thruster Assembly. Figure 8.28 shows the opposing side of the Spacecraft with the 4 body mounted solar array panels.

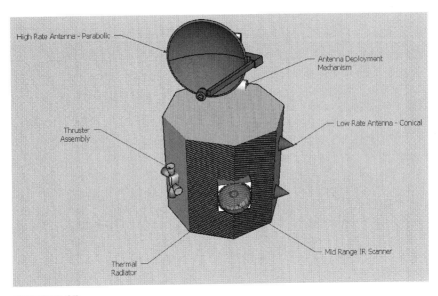

FIGURE 8.27
Conceptual View of the Spacecraft Geometry.

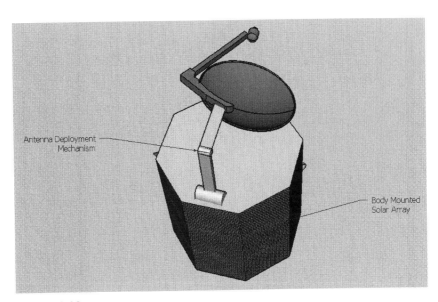

FIGURE 8.28
Conceptual View of the Spacecraft Geometry - Opposing Side.

The consistency between the system model and the geometric model must be maintained to ensure the integrity of the system design. The system model is generally a more abstract model that captures the system breakdown to the component level, but not down to the piece part level (e.g., nuts and bolts). The components, such as a circuit card or mechanical component, can be identified in the system model, and mapped to a corresponding component in the system bill of materials (BOM) or master equipment list (MEL). Third party tools can be used to maintain the mapping between the system model and the BOM, which in turn can be related to the components and assemblies in the geometric model.

The model configuration management environment is used to capture versions of the preferred system design configuration as it evolves, while maintaining the design history and rationale that led to this selection in previous versions of the model. As the design continues to evolve, the system model must be maintained to ensure consistency between the architecture and design. Later in the lifecycle, the component blocks with their critical properties can be specialized and updated to reflect the verification results from integration and test activities. The maintenance of this model requires an established infrastructure and the application of a disciplined change management process.

PERFORM DESIGN ANALYSIS

CHAPTER 9

The purpose for this activity is to obtain quantifiable results to support trade studies among design alternatives, determine parameter sensitivities to optimize a design, and evaluating requirements satisfaction. The Perform Analysis activity continues throughout the development process. Some examples of analysis are referred to in earlier chapters, such as the orbit analysis in Chapter 6.4, the mass and delta-v analysis in Chapter 7.3, and the Solar Array trade study analysis in Chapter 8.2.7.

This chapter addresses the following:

- Identify the mission and system analysis to be performed
- Specify each analysis in terms of its input and output parameters and their relationship to the system design properties
- Perform the engineering analysis using the appropriate engineering analysis tool to determine the values of system properties.

9.1 Define the Analysis Context

Some of the analyses that are performed at the mission level are identified in Figure 9.1. Each analysis is intended to analyze a particular measure of effectiveness (MoE) that contributes to the overall cost-effectiveness of the solution. The MoE's were previously identified in the top level block definition diagram in Figure 6.6. The Analysis Context block aggregates each analysis, and references the Mission Context as the subject of the analysis. This pattern supports the ability to reuse analytical models to analyze different subjects of analysis in different mission contexts.

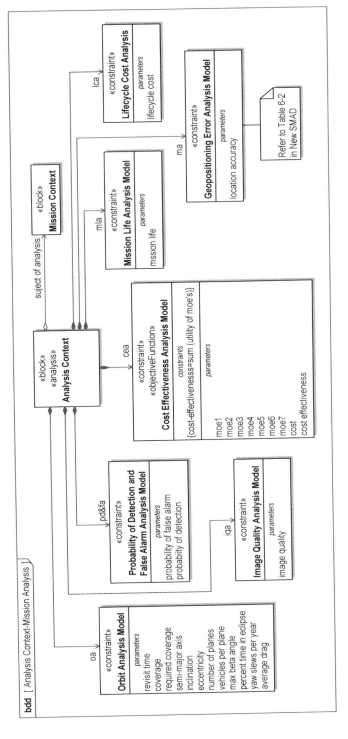

FIGURE 9.1

This Analysis Context aggregates the constraint blocks corresponding to different mission analysis models used to analyze the measures of effectiveness.

9.2 Specify and Perform the Analysis

The results of executing the Orbit Analysis Model are summarized in Figure 6.8 as part of the Analyze Mission & Stakeholder Needs activity. The analysis tool that executes the Orbit Analysis Model is the AGI System Toolkit (STK) that is highlighted in Figure 6.7. The Orbit Analysis Model constraint in Figure 6.8 is a proxy for the actual analytical model that is executed by the analysis tool. The parameters of the constraint correspond to the inputs and outputs to and from the analytical model.

The corresponding parametric diagram is shown in Figure 9.2. The parametric diagram includes the constraint called Orbit Analysis Model. This constraint includes input and output parameters to the analytical model which are bound to the properties of the design. This constraint serves as a specification of the analytical model that executes the detailed equations to derive the optimal orbit. The input-output direction is not shown in the figure since a parameter may be treated as a dependent or independent parameter in different contexts(i.e. $y=k*x$ or $x=y/k$).

The revisit time and coverage are primary output parameters from this analysis. Critical input parameters include orbit eccentricity, inclination, semi-major axis, number of planes, and number of vehicles per plane. This specific analysis is meant to represent a coarse grained, constellation design exploration model. As such, not ever orbital element (e.g. RAAN) is passed to the model as these are iterated on internally to the analysis model. Other orbit parameters include max beta angle to the sun and percent time in eclipse, which are also input to the Solar Array sizing analysis.

As noted previously, the parametric diagram binds the critical input and output parameters of the orbit analysis constraint to properties of the system design model. For example, the eccentricity parameter of the Orbit Analysis Model is linked to the eccentricity property of the Orbit. Similarly, the revisit time parameter of the Orbit Analysis Model is linked to the revisit time property of the Mission Enterprise. In this example, the names of the properties are the same as the parameter names of the analysis model, but they are not required to be the same. In particular, this same Orbit Analysis Model can be reused to analyze different system design alternatives, each of which may have different property names.

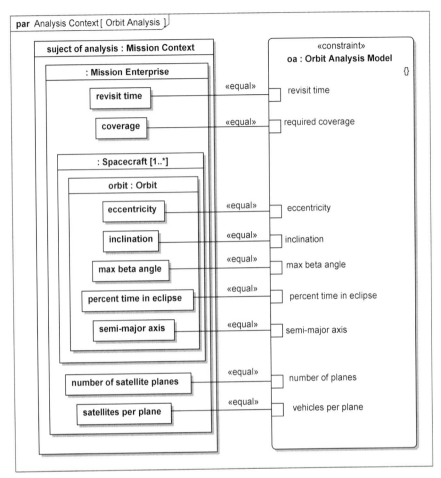

FIGURE 9.2
A SysML Parametric diagram that specifies the orbit analysis for the Spacecraft.

The Analysis Context for some of the system level analysis models is shown in Figure 9.3. This contrasts with the mission level analysis models in Figure 9.1.

ARCHITECTING SPACECRAFT WITH SYSML

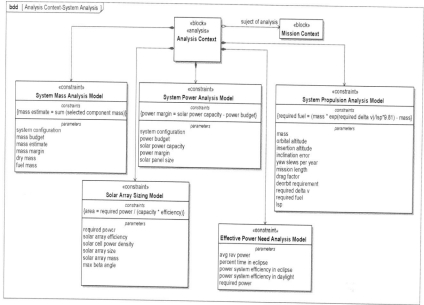

FIGURE 9.3

The additional systems analysis models are shown to support power analysis, mass analysis, and propulsion analysis.

The Mass and Delta-V analysis is used to derive the Mass and Delta-V requirements in Figure 7.7. The parametric diagram in Figure 9.4 specifies the analytical models needed to support the mass and delta-v analysis with the system design. In this example, four analyses are required to determine the required fuel mass for the Spacecraft. The four analysis include the orbit analysis discussed previously, a delta-v (required velocity change) analysis to determine requirements on the propulsion system, the mass analysis to determine the system dry mass, and the propulsion analysis to determine the required fuel mass. Some of the parameters, such as the mass parameter of the mass analysis model, are bound to other analysis models, such as the mass parameter of the propulsion analysis model, to provide an integrated view of the analysis models.

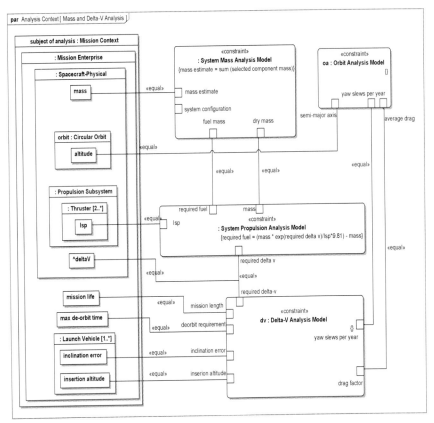

FIGURE 9.4

The parametric diagram for Mass and Delta-V analysis includes the mass analysis, propulsion analysis, and orbit analysis.

The parametric diagram for the Solar Array trade study discussed in Chapter 8.2.7 is shown in Figure 9.5. This trade study includes analysis models for power, mass, reliability, and cost. In this particular case, the analysis models are executed using Microsoft Excel®, but more generally, each analysis model can be executed by a different tool.

ARCHITECTING SPACECRAFT WITH SYSML

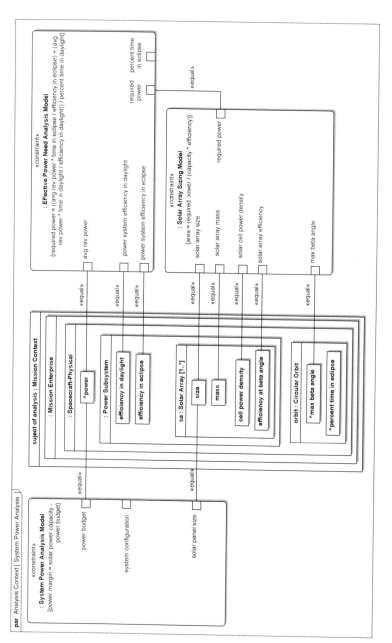

FIGURE 9.5

Power system sizing and analysis for the Solar Array trade-study.

CHAPTER 10

MANAGE REQUIREMENTS TRACEABILITY

This purpose for this activity is to establish traceability relationships to support requirements flow down and change impact analysis. The Manage Requirements Traceability activity continues throughout the development process. There are several different kinds of relationships that are used to relate requirements, design elements, analysis, and verification. Some of these relationships between elements at the mission, system, and component levels are illustrated in the previous chapters. This chapter describes how traceability relationships are used to ensure the design satisfies its requirements, and to assess the impact of changes to requirements and design.

This chapter addresses the following:

- Define the specification hierarchy.
- Define the requirements contained in each specification
- Establish traceability relationships that include:
    o Requirements to requirement (e.g., derive)
    o Requirements to design (e.g., satisfy)
    o Requirements to test case (e.g., verify)
- Assess traceability and identify gaps

10.1 Define the Specifications

The mission, system, and component requirements are captured as shown in Figure 6.4, Figure 7.6, and Figure 8.24 respectively. Each specification contains the text requirements that specify an element at the mission, system, subsystem, or component level of design. For example, the mission requirements specify the Mission Enterprise, the system requirements specify the Spacecraft, and the component requirements specify a particular hardware or software component, which is the Payload Sensor in this example.

The Specification Tree in Figure 10.1 establishes a trace relationship between each of the specifications at the mission, system, and component levels. In this example, the original source Mission Requirements from the New SMAD are at the top of the tree, and the Mission Requirements-Refined trace to these source requirements. The Spacecraft Specification traces to the Mission Requirements-Refined, and the Payload Sensor Specification is a component specification that traces to the Spacecraft Specification. Additional subsystem and hardware and software component specifications can be captured in a similar way to provide a more complete view of the Spacecraft Specification Tree.

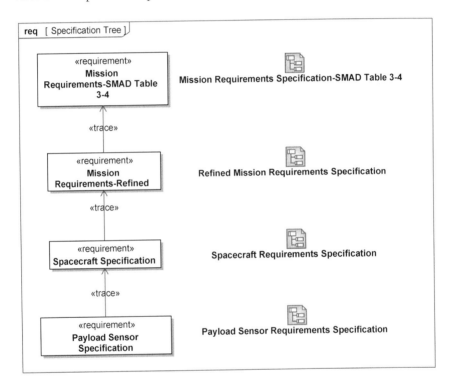

FIGURE 10.1

An example of trace relationships between specifications used to represent the specification.

10.2 Establishing Traceability

The traceability relationships are established throughout the development process, such as the relationships used to derive the orbit requirements in Figure 6.9, the relationships used to derive the mass and delta-v requirements in Figure 7.7, and the mass satisfaction relationship in Figure 8.26. An example of how critical Payload Sensor requirements trace to the mission requirements is shown in Figure 10.2. In this example, the key mission functional requirement to Detect and Monitor Forest Fires in US and Canada. The Detection Time is a critical mission performance requirement associated with the functional requirement. The required Spacecraft orbit Revisit Time to a particular region on earth and the Probability of Detection requirements are derived from the Detection Time requirement. The Payload Sensor functional requirements and the performance requirements for resolution and sensitivity are further derived from the probability of detection and the orbit altitude requirements. The rationale for the derivation is based on the Sensor Performance Analysis. These Payload Sensor functional and performance requirements are satisfied by the Mid-Range IR Scanner resolution and sensitivity properties and its function to sense thermal emissions, as indicated by the satisfy relationships. Although not included in this example, the more complete traceability includes relationships between the Payload Sensor requirements and other Spacecraft requirements, such as Pointing Accuracy. These relationships can be navigated to assess the impact of requirements and design changes.

The test cases used to verify the Payload Sensor satisfies its resolution and sensitivity requirements are also shown in the figure. Elaboration of the test cases that defines how verification is accomplished is discussed in the next chapter.

The traceability relationships shown in Figure 10.2 can be presented in tabular and matrix form to provide a more scalable approach to managing large numbers of traceability relationships. The management of the requirements relationships can also be augmented by other requirements management tools that can then be synchronized with the model.

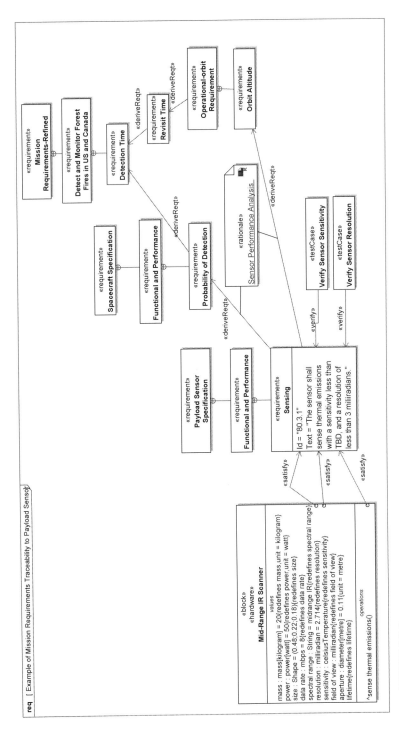

FIGURE 10.2

The traceability relationships from Payload Sensor component to the mission requirements includes relationships between requirements, design, analysis, and verification elements, and can be used to support change assessments.

VERIFY SYSTEM

CHAPTER 11

This purpose of this activity is to establish the approach to verify that the requirements are satisfied. This process should begin with the development of the requirements to ensure the requirements are verifiable in a cost-effective way. The verification methods typically include a combination of inspection, analysis, demonstration and test. The model can be used to define the verification method. When integrated with analysis tools, the verification method can be executed to verify the system by analysis. This chapter provides a brief introduction to how the model can be used to support verification planning, and integrate verification with the system requirements, design, and analysis.

This chapter addresses the following:

- Define the verification context that includes the verification environment and the unit under verification
- Elaborate the test cases to specify how the requirements are verified
- Define the sequence of test cases to support test planning

11.1 Define the Verification Context

The test cases in Figure 10.2 are used to verify that the Payload Sensor sensitivity and resolution requirements are satisfied. The Verification Context in Figure 11.1 brings together the elements needed to perform the test cases to achieve the verification objectives. For these Payload Sensor test cases, the verification context includes the Mid-Range IR Scanner as the unit-under-test, and the test environment which includes the Test Operator and the test components (i.e. test equipment). The test components are an Optical Bench, a Signal Processor, and a Display.

An internal block diagram for the verification context can define the interconnection between the test components and the unit-under-test (not shown).

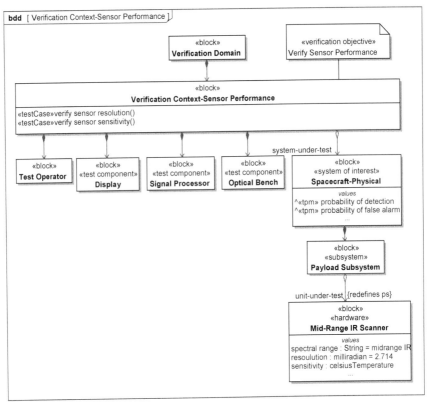

FIGURE 11.1

The verification context defines the elements needed to perform the test cases to achieve the verification objectives. This includes the unit-under-test (i.e., the Mid-Range IR Scanner), and the verification environment, which includes the Test Operator and the test components.

11.2 Elaborate the Test Cases

The test cases specify the actions to achieve the verification objectives and associated pass/fail criteria. These actions can be defined at a high level or at the detailed level of a test procedure depending on the need. Each test case is intended to verify one or more requirements as identified by the verify relationships between the test cases and the requirements. The test case can be represented by an activity diagram or other behavior to show the actions performed by the Test Operator, test components, and unit-under-test to verify the requirement is satisfied. An example is shown in the activity diagram to Verify Sensor Resolution in Figure 11.2.

The Test Operator controls the unit-under-test and the Optical Bench including its thermal source to generate the needed stimulus. The Mid-Range IR Scanner produces a sensor output which is then processed by a Signal Processor and displayed by the Display to the Test Operator. The Test Operator measures the resolution and assesses whether the Payload Sensor satisfies its resolution requirement. Other test cases are elaborated in a similar way to verify the requirements.

The model of the end-to-end system or enterprise may include a combination of simulated components, operational hardware, software, and people in the loop, or the complete as-built end-to-end system configuration. The elements of the test case represent design elements early on, but may represent actual as-built elements later in the lifecycle. For example, the unit-under-test in the model may initially represent the design of the Mid-Range IR Scanner, but later on represent an actual Mid-Range IR Scanner in a test-bed with its measured values. Eventually, the model can represent the actual spacecraft in a test lab, with a simulated ground station, and the end user viewing the results on a representative display.

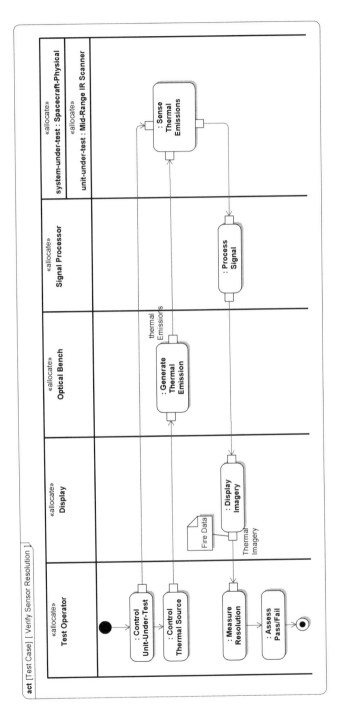

FIGURE 11.2

The test case is elaborated as an activity diagram to show the actions performed to verify the Payload Sensor resolution requirement.

11.3 Define the Test Sequence

The activity diagram in Figure 11.3 shows the sequence of test cases that are used to verify the Payload Sensor requirements. This information, along with the test context that describes the verification objectives, test components, and test configurations, and the verification relationships with the requirements being verified, provide critical inputs for test planning.

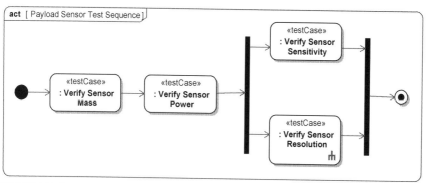

FIGURE 11.3

The sequence of test cases to verify the Payload Sensor requirements is shown in an activity diagram.

SUMMARY

CHAPTER 12

This book describes a model-based systems engineering (MBSE) approach that is applied to the design of a small spacecraft. The Spacecraft design is inspired by the FireSat II example in the New SMAD. The intent of the example is to demonstrate how this approach can support the mission and system specification, design, analysis, and verification of complex systems, and add value over traditional document-based approaches. The approach also highlights how the model can be used as a reference architecture that can be realized by multiple alternative designs that are subject to trade study, and can support a range of mission needs.

The fundamental tenet of this approach is to capture this critical information in a robust model of the system (i.e., the system model) that can be viewed from many different perspectives. The system model is maintained throughout the system life cycle as a fundamental part of the technical baseline.

The key enablers of this approach are a modeling language, a well-defined MBSE method, and a system modeling tool. In order to fully leverage this approach, it should be part of an integrated model-based engineering approach that includes integration of the methods, tools, and models across engineering disciplines. The MBSE method with SysML described in this book provides a starting point for applying an MBSE approach.

MBSE can enable good systems engineering, but still demands competent systems engineers with the essential domain knowledge to apply a rigorous engineering approach, and an understanding of how to apply the modeling language, MBSE method, and modeling tool, each of which has its own complexities. It is important to remember that the model is only as good as the data that goes into it, and should be subject to on-going peer review by domain experts.

Implementing a successful MBSE approach in an organization requires an investment in the tools, training, and skill development, and an on-going commitment to learn from the experiences to improve the practice. This should be considered as part of a longer term strategy and an incremental improvement process to meet the complexity, quality, and productivity challenges facing system development. It is anticipated that the MBSE methods along with the system modeling languages and tools will continue to evolve to meet these challenges for performing effective systems engineering.

REFERENCES

APPENDIX A

[1] Advances in Systems Engineering edited by John Hsu and Richard Curran, American Institute of Aeronautics and Astronautics, 2016, Chapter 4 Applying SysML and a Model-Based Systems Engineering Approach to a Small Satellite Design

[2] Space Mission Engineering: The New SMAD edited by James R. Wertz, David F. Everett, and Jeffery J. Puschell.

[3] CubeSat Mission Design Based on a Systems Engineering Approach Sharan A. Asundi, and Norman G.Fitz-Coy, IEEE Aerospace Conference, 2013

[4] MBE NDIA Report: Final Report of the Model Based Engineering (MBE) Subcommittee, NDIA Systems Engineering Division, M&S Committee, 10 February 2011

[5] DoD. 1998: "'DoD Modeling and Simulation (M&S) Glossary" in DoD Manual 5000.59-M. Arlington, VA, USA: US Department of Defense. January. P2.13.22. Available at http://www.dtic.mil/whs/directives/corres/pdf/500059m.pdf

[6] Object Management Group. OMG Systems Modeling Language™ (SysML®). V1.5. Available at: http://www.omg.org/spec/SysML/.

[7] ISO/IEC 19514:2017, Information technology—Object management group systems modeling language (OMG SysML). Int Organ Standardization/Int Electrotechnical Comm May 2017

[8] A Practical Guide to SysML, The Systems Modeling Language, Third Edition by Sanford Friedenthal, Alan Moore, and Rick Steiner, Morgan Kaufmann, 2015

[9] SysML Distilled, A Brief Guide to the Systems Modeling Language, Lenny Delligatti, Addison Wesley, 2013

[10] ISO/IEC 15288:2008, Systems and Software Engineering—System Life Cycle Processes. Int Organ Standardization/Int Electrotechnical Comm March 18, 2008;.

[11] Model Based Systems Engineering On The Europa Mission Concept Study, IEEE Aerospace Conference Proceedings, 2012, T. Bayer, S. Chung, B. Cole, B. Cooke, F. Dekens, C. Delp, I. Gontijo, K. Lewis, M. Moshir, R. Rasmussen, D. Wagner.

[12] Update on the Model Based Systems Engineering on the Europa Mission Concept Study, IEEE Aerospace Conference Proceedings, 2013, T. Bayer, S. Chung, B. Cole, B. Cooke, F. Dekens, C. Delp, I. Gontijo, D. Wagner.

[13] A Representative Application of a Layered Interface Modeling Pattern, Sanford Friedenthal, Marc A Sarrel, Peter M Shames, 26th Annual INCOSE International Symposium (IS 2016) Edinburgh, Scotland, UK, July 18-21, 2016

[14] Pegasus Launch Vehicle Users Guide. Available at: http://www.orbital.com/LaunchSystems/Publications/Pegasus_UsersGuide.pdf April 2010, Release 7.0

APPENDIX B

ABOUT THE AUTHORS

Sanford Friedenthal is an industry leader in model-based systems engineering (MBSE) and an independent consultant. In his previous position as a Lockheed Martin Fellow, he led the corporate engineering effort to enable Model-Based Systems Development (MBSD) and other advanced practices. In this capacity, he was responsible for developing and implementing strategies to institutionalize the practice of MBSD across the company and providing direct model-based systems engineering support to multiple programs.

His experience includes the application of systems engineering throughout the system lifecycle from conceptual design through development and production on a broad range of systems. He has also been a systems engineering department manager responsible for ensuring that systems engineering is implemented on programs.

He also was a leader of the industry team that developed SysML from its inception through its adoption by the OMG. Mr. Friedenthal also led the International Council on Systems Engineering (INCOSE) MBSE Initiative, and is co-author of 'A Practical Guide to SysML'.

Christopher Oster is a Systems and Software Architect at Lockheed Martin Space Systems and an industry recognized expert in Systems Engineering and model-integration. Currently, Mr. Oster leads efforts related to next generation command and control systems and data analytics efforts. In his prior role, Mr. Oster was the Lockheed Martin Commercial Space Model-based Enterprise lead where he was responsible for ensuring the successful transition of Lockheed Martin Space portfolio to a model-based engineering methodology.

Mr. Oster's experience has focused on developing large scale, software intensive systems including data processing suites, spacecraft command and control systems and space platforms. At Lockheed Martin since 2004, Mr. Oster has held roles in software development, systems engineering, architecture, business development and R&D, and is a graduate of Lockheed Martin's Advanced Technical Leadership Development program. Mr. Oster is also an active member of the systems engineering and model-based communities, as a frequent contributor to the INCOSE MBSE initiative, a member of the INCOSE Systems 2025 Vision project and frequently lectures at universities, industry conferences and government offices about model-based systems engineering. Prior to joining Lockheed Martin, Mr. Oster earned his M.S. and B.S. in Computer Science and Engineering from The Pennsylvania State University in State College, Pennsylvania and is currently enrolled in a PhD program in Systems Engineering at Stevens Institute of Technology.

Made in the USA
Middletown, DE
01 November 2023

41786429R00077